GENERAL CHEMISTRY I LAB | CHEM 121L

Louisiana State University Shreveport
Department of Chemistry and Physics

Revised March 2020

General Chemistry I Lab

CHEM 121L
Louisiana State University Shreveport
Department of Chemistry and Physics
Revised March 2020

Printed in the United States of America
10 9 8 7 6 5 4 3 2 1
ISBN: 978-1-61740-951-6

Van-Griner Learning
Cincinnati, Ohio
www.van-griner.com

President: Dreis Van Landuyt
Project Manager: Janelle Lange
Customer Care Lead: Lauren Houseworth

Yu 951-6 Su20
318380
Copyright © 2021

Edited by

William W. Yu, Wayne A. Gustavson

Revision Notes

This lab manual was revised from the 2002 version with 13 experiments, initially edited by Dr. Gustavson and colleagues.

03/2014   Not printed. Added instructions on Excel and graduated cylinder, included all needed information, put all things in one manual, determined the formatting style.

05/2014   Printed. Incorporated 2 new experiments (15 experiments).

12/2014   Printed. Corrected language problems, modified the description for Molar Volume of Gases experiment.

03/2017   Not printed. Changed some descriptions.

05/2018   Printed. Rearranged a few experimental procedures, reduced and adjusted some chemical quantities, modified some descriptions.

08/2018   Not printed. Rewrote several after–lab questions.

03/2020   Printed. Amended some discussion and descriptions, cited new references.

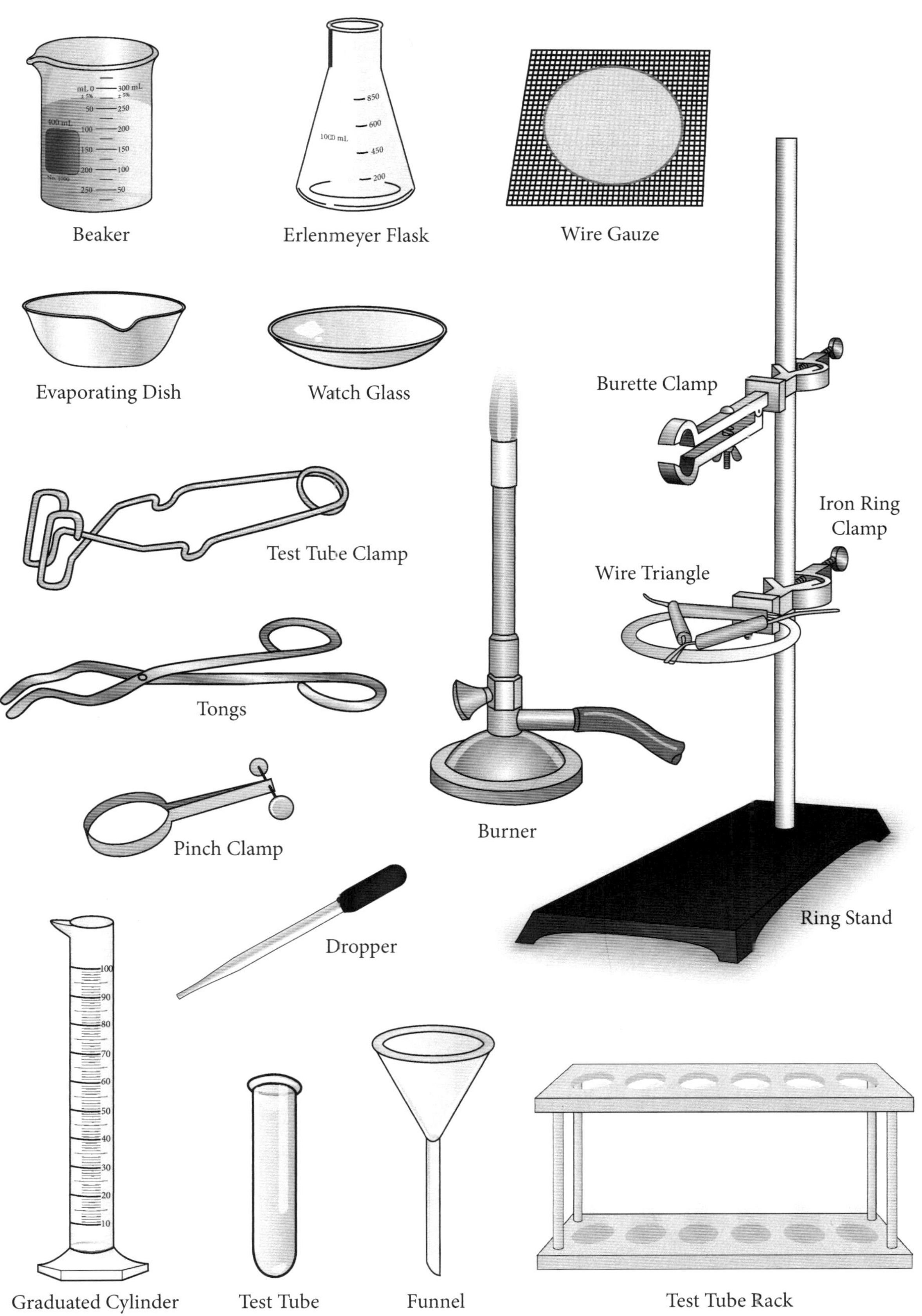

Labortory Equipment

Table of Contents

Laboratory Safety

The laboratory is a place for the experimental study of science and its applications. A quiet, orderly, and safe atmosphere is essential to achieve the best results. Your safety in the laboratory is insured if you and your classmates know and follow fundamental safety rules. These rules are extremely important and must be carefully read and understood.

1. Locate all the safety features in your laboratory such as the hoods, eye wash, fire extinguisher, fire blanket, and safety shower so that you can use them if needed. In case of fire or accident, call the instructor at once.

2. Perform only authorized experiments. Unauthorized "experimenting" is dangerous to everyone and may lead to your dismissal from the class.

3. Wear proper clothing. To protect the legs and feet from chemical spills, wear full−length pants and shoes (not sandals) whenever possible in the lab. Also, it is advised that old clothes be worn as minor chemical spotting will occur from time−to−time.

4. Confine long hairs at all times.

5. Skin contact with chemicals is the most common form of accident, usually affecting the hands and arms. The key word for treatment is water, large volumes of it. Immediately wash the affected area with a large amount of water. Do this even if no sensation of burning is felt since many chemicals are slow acting.

6. Wear approved eye protection in the laboratory continuously. You are responsible for obtaining a pair of **safety goggles**. Prescription glasses or sports goggles are not considered sufficient for this laboratory course. Contact lenses should not be worn in the laboratory.

7. If you should get a chemical in your eye, wash with flowing water from the safety eye wash for 10 ~ 15 minutes. Report the accident to your instructor immediately and see a physician. Regardless of how minor you may think they are, report all accidents immediately to the instructor.

8. Do not taste anything in the laboratory. This applies to food as well as chemicals. Do not use the laboratory as an eating place and do not eat or drink from laboratory glassware.

9. Avoid breathing fumes of any kind. Perform experiments involving toxic fumes under the hood. If you must smell a chemical, use an indirect wafting technique.

10. Handle all chemicals with the proper equipment (e.g., spatula, pipet, beaker, etc.). Never use mouth suction in filling pipettes; use a pipet bulb. Do not use cracked or chipped glassware. Obtain suitable replacements from the stockroom.

11. Never force glass tubing into rubber stoppers. Always lubricate the tubing with water or glycerine and protect your hands with a cloth towel. Hold hands close together to minimize the leverage on the glass.

12. Never work in the laboratory alone.

13. Never leave a burner flame or reaction unattended.

14. Keep laboratory benches and desk drawer clean and neat.

15. Any spills no matter how small on the bench or floor must be taken care of immediately to prevent others from unknowingly getting injured by them. Check with the instructor for cleanup procedures. **Acid spills** should be neutralized with sodium bicarbonate ($NaHCO_3$), then diluted with water, and the area cleaned up. **Base spills** should be neutralized

with boric acid (H_3BO_3), then diluted with water, and the area cleaned up.

16. Disposal of chemicals should be done carefully. If you are not sure about the proper disposal of any chemical and waste, ask your instructor.

17. Allow items that you obtained from the common drawer (such as a ring stand or a burner) to **cool sufficiently** before returning them to their original location.

18. Before you leave the laboratory make sure that the gas and water are turned off, that all bottles are in place, and that your desk top is **clean**.

Become "safety conscious"! Developing a respect for chemicals will serve you well in the laboratory as well as in everyday life.

Laboratory Notebook

In each laboratory course in chemistry, each student will be required to maintain a notebook that will be the permanent record of all the work performed in the laboratory. Scientific experimentation should always be accompanied by such a permanent record that provides evidence that the work was done as well as the opportunity for it to be repeated by others. Keep in mind that a properly maintained notebook would alone allow another student to repeat your experiment. The instructions listed below are to be followed in organizing your notebook.

1. The student's name, the course, and the laboratory section should be written clearly on the outside cover of a **bound** notebook.
2. The first two or three pages should be left blank initially for later use as a Table of Contents of the laboratory experiments.
3. The pages thereafter should be numbered consecutively. Do not remove any pages from the notebook.
4. The entire record must be kept in **ink**.
5. Erasures are not allowed in the notebook. If a mistake is made, neatly draw a single line through the error and insert the correction. For example,

the ~~flask~~ was
beaker

6. The experiments are to be entered in the notebook in the sequence in which they are performed in the laboratory. No pages should be left empty in the event an experiment was missed.

7. In order to prepare a good notebook, you must study each experiment before coming into the laboratory. As evidence of such preparation, you should have included on a new page the following information:

1) **TITLE** of the experiment. Use the title shown in the laboratory manual.

2) **OBJECTIVE** (or **PURPOSE**) of the experiment. Be brief.

3) **EXPERIMENTAL PROCEDURES**. This section is to be completed in the notebook before the class begins. Procedures will be taken from the laboratory manual in some cases with modifications made by the instructor. Write exactly what is to be done, what chemicals or pieces of equipment are to be used, how they will be treated, etc. Include all significant details so that it will indeed show exactly what you will do. A complete record will enable you or your instructor to discover the reason for an unexpected or incorrect result.

4) **DATA and RESULTS**. Record **all data** (such as weight, volume, temperature, length, reaction time) and note all results and pertinent observations (color change, temperature difference, etc.) in your notebook as they are determined. Numerical data must be recorded using the proper number of significant digits including the unit of measurement (cm, g, mL, etc.). Do not record information on scraps of paper because these are easily lost! Although the notebook should be as neat as possible, neatness is not the primary objective. A laboratory report is the medium used for neatly summarizing an experiment. Thus, the student should not hesitate to record all items directly in the notebook even if it does not seem completely organized at the moment. Advanced planning and experience will allow you to develop better organization of this section.

5) **DISCUSSION**. This final section of the record is a short discussion of the experiment including any significant conclusions, error or mistake analyses.

6) Each page of the experimental record should be dated, preferably in the upper left–hand corner. Sign your name at the end of the experiment or at the point where work was discontinued for the day. When work is resumed, enter the new date at that point on the page and continue. Any space remaining at the end of an experiment should be left blank. Late additions or corrections to completed and signed materials should be so indicated, dated, and signed.

7) When extra charts or graphs are prepared, they should be **permanently fastened** into the notebook (using glue, staples, or tape) following the final page for that experiment. Be sure to title each chart or graph and label all axes or columns of data.

8) Leave a page blank at the end of each experiment. When the laboratory report is graded and returned, permanently fasten it into the notebook on this blank page.

With these rules carefully followed, a laboratory report can be easily assembled. Furthermore, the notebook record will now be in such a condition that the student will find it easy to derive meaning from the experiment upon review and therefore be in a position to understand better the scientific principles that are illustrated by the experiment.

The notebook is subject to inspection by the instructor at any time. It thus behooves the student to come to the laboratory prepared as indicated above.

Basic Concepts and Techniques

We will discuss several basic concepts and techniques for General Chemistry I laboratory course. These are:

1. Error Analysis
2. Significant Figures
3. Graphical Analysis by Excel Spreadsheets
4. Use of Analytical Balances
5. Use of Triple Beam Balances
6. Use of Top−loading Balances
7. How to Read Graduated Cylinders

1. Error Analysis

An understanding of the terms, procedures, and interpretations involved during the collection of experimental data enables a scientist to evaluate data which otherwise might be meaningless. The following terms are designed to help the student understand some of the factors involved in error analysis.

1) Determinant Errors – often called Systematic Errors – those errors for which corrections can be made.

2) Random Errors – encountered in all measurements, are beyond the control of the observer. These errors are normal fluctuations of observed quantities.

3) Precision – the reproducibility of the measurement. If the same measurement is made repeatedly, the precision deals with how close the measurements are to one another. Good precision means that the measurements are all close together. The values may be incorrect, but they are closely grouped.

4) Accuracy – deals with the difference between a measured quantity and the actual value. In the laboratory, data needs to be precise as well as accurate.

5) Mean Value – often called an Average – the sum of the measured quantities divided by the number of measurements.

6) Deviation – the absolute value of the difference between a measured quantity and the mean value.

7) Mean Deviation – the mean (or average) of the deviations. This quantity is calculated by adding together the deviations of each of the values and then dividing by the number of measurements that were taken.

8) Relative Deviation – this quantity is calculated by dividing the mean deviation by the mean value. It may be expressed as a percent (%) by multiplying by 100, or as parts per thousand (‰) by multiplying by 1000.

An example is given for obtaining the mean value, deviation, and mean deviation.

Observed values	Deviations
1.014	$\mid 1.014 - 1.013 \mid = 0.001$
1.011	$\mid 1.011 - 1.013 \mid = 0.002$
1.013	$\mid 1.013 - 1.013 \mid = 0.000$
	Sum of deviations = 0.003
Sum = 3.038	Mean deviation = 0.003/3 = 0.001
Mean value = 3.038/3 = 1.013	Relative deviation = (0.001/1.013) × 100% = 0.1%

An outlaying observation may be dropped if its deviation from the mean is greater than 4 times the mean deviation of the other observed values. This rejection is valid provided at least 4 observations, including the suspicious one, have been made.

2. Significant Figures

In science it is important to relate the accuracy of the data that is collected in the laboratory. Significant figures are the meaningful digits in a measured quantity. For example, the volume of a solution is between 40 and 41 mL. The graduated cylinder has 1 mL intervals, so you estimate the volume to the nearest 0.5 mL, and record a volume of 40.5 mL. This means that there is some error in the last recorded digit, but the volume is more accurate than either 40 mL or 41 mL. The rules for the number of significant figures in a measured value are given below.

1) All nonzero digits are significant.

    462 – 3 sig figs; 1.2438 – 5 sig figs

2) Zeroes placed between nonzero digits are significant.

    10.3 – 3 sig figs; 10.407 – 5 sig figs

3) Zeroes at the right of the last nonzero digit and to the right of the decimal point are significant.

    14.100 – 5 sig figs; 1.53670 – 6 sig figs

4) Zeroes to the left of the first nonzero digit are **NOT** significant – these zeroes function only to locate the decimal point.

    0.0025 – 2 sig figs; 0.1090 – 4 sig figs

5) Addition and Subtraction of Significant Figures – a sum or difference cannot be more accurate than the least accurate number involved.

Add the following values: $75.1 + 1.005 + 1.4537 = 77.5587$

The least accurate value involved in the addition is 75.1 because it is known to only the tenths place. This means that the answer can be known to only the tenths place as well, so the answer to the correct value of 3 significant figures is 77.6.

6) Multiplication and Division of Significant Figures – the number of significant figures in the answer is the same as the fewest significant figures in the calculation.

Multiply the following: $4.18 \times 1.6324 = 6.8234$

4.18 has 3 sig figs, and 1.6324 has 5 sig figs. This means that the answer can only have 3 sig figs, and is therefore reported as 6.82.

3. Graphical Analysis by Excel Spreadsheets

The calculations and data analyses in experiments can be simplified by the use of a spreadsheet. Spreadsheets are especially useful when large numbers or repetitive calculations are necessary. The spreadsheet program that will be discussed here is Excel. Excel is a key component of Microsoft Office, and is available on the computers in SC 311 (Computer Lab). A general use of Excel is described here, and more specific instructions will be provided with the experiments.

1) Open Excel (2010 version is shown here) to have a blank file. Type the data into the Excel sheet 1. You can type the data in any columns, we assume you use A and B columns (the column width is adjustable just

like a table column in Word; a simple way is to double click the right side of that column) (Figure 1a).

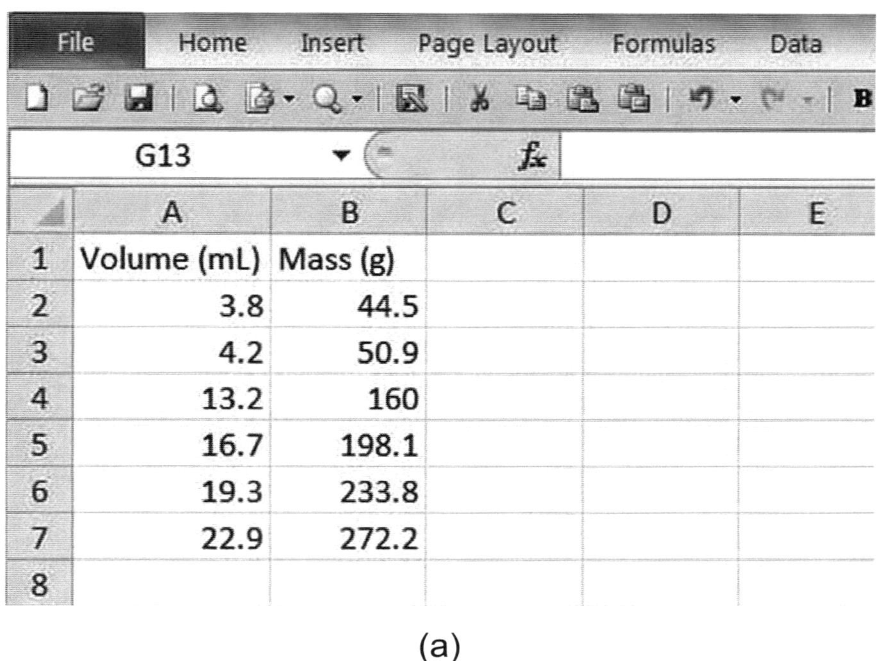

(a)

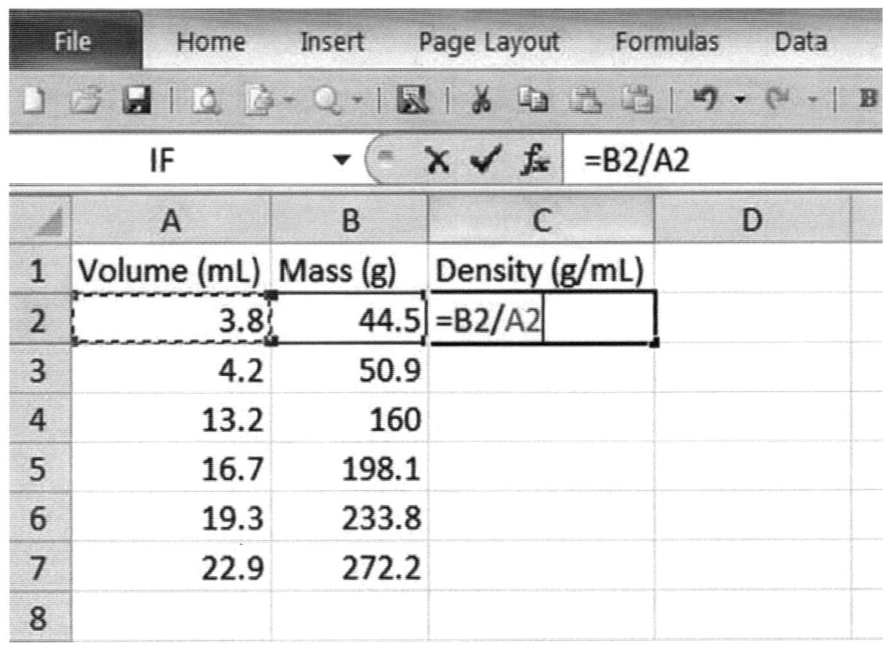

(b)

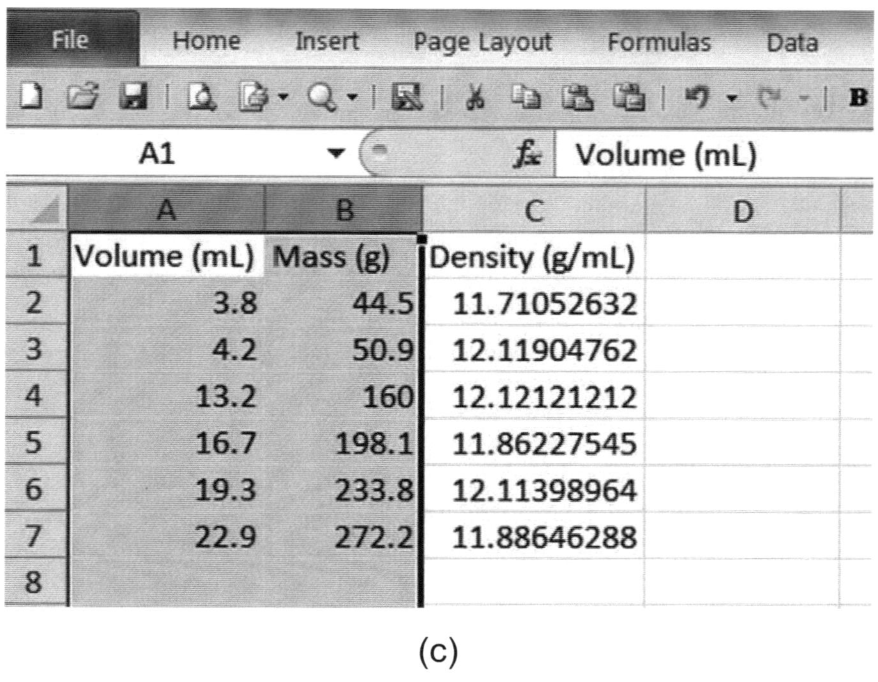

(c)

Figure 1. Input data and calculate new data.

2) When calculations are required, the formula used for the calculation must be entered at the top of the column. Calculations are typed into the spreadsheet as mathematical functions. They must be proceeded by " = " to tell Excel to perform a calculation. If, for example, you want to multiply the value in A2 by the value in B2 and then divide by the value in C2, you must enter the following formula: "=(A2*B2)/C2". Figure 1b shows the calculation of "=B2/A2" which gives the density.

3) Mathematical functions are represented by " + " for addition, " − " for subtraction, " * " for multiplication, and " / " for division. If calculation needs to be repeated for each of the values in a column, the equation need only be typed in once. To perform this task easily, use the mouse to highlight (double click) the box in which the calculation has already been performed. Click the right bottom of that box and hold down the

left mouse button; move the cursor down to the last box in the column, and then release. This should calculate the values for all of the values in the column (Figure 1c).

4) Follow the instructions to plot the Mass–Volume graph and find the slope, intercept, and the R^2 value of the plot.

a) Highlight both columns of data (Figure 1c);

b) Go to "insert", choose "scatter", and "scatter with only markers" (Figure 2a);

c) To add a regression line, click on any of the data points;

d) Then right click the mouse and select "add trendline".

e) Choose "linear" and check the boxes next to "display equation on chart" and "display R squared value on chart" (at the bottom of the "add trendline" window);

f) Upon completing this step, an equation in the form of $Y = a\,X + b$ will be displayed on the graph along with a R^2 value. In this equation, a is the slope of the straight line, b is the Y intercept (Figure 2b).

g) Format the equation so that the slope and the intercept of the equation have at least 3 significant figures. This is necessary to minimize rounding error. This step is more important when the values of the slope and the intercept are small (< 0.01). To format the equation, click on the equation, right click the mouse and select "format trendline label".

h) Everything on the final graph (Figure 2b) is editable. Label the axes and add a chart title to the graph.

i) The final graph can be copied to your Word lab report. See instructor for help if you have questions about this step.

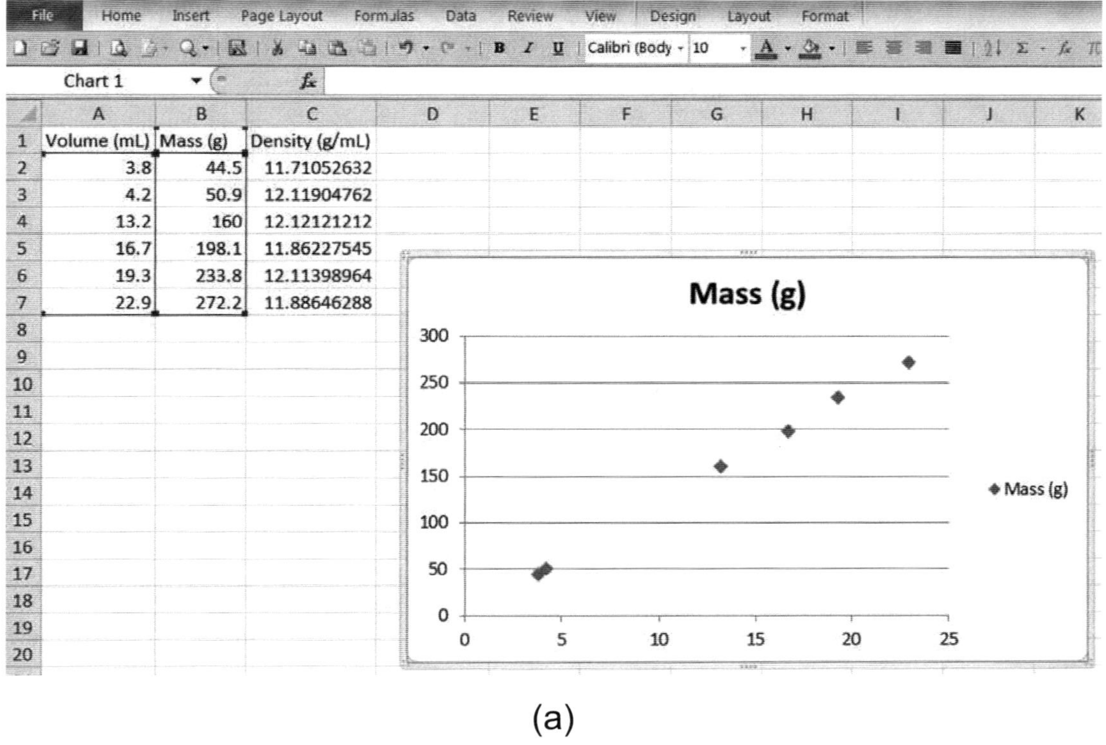

(a)

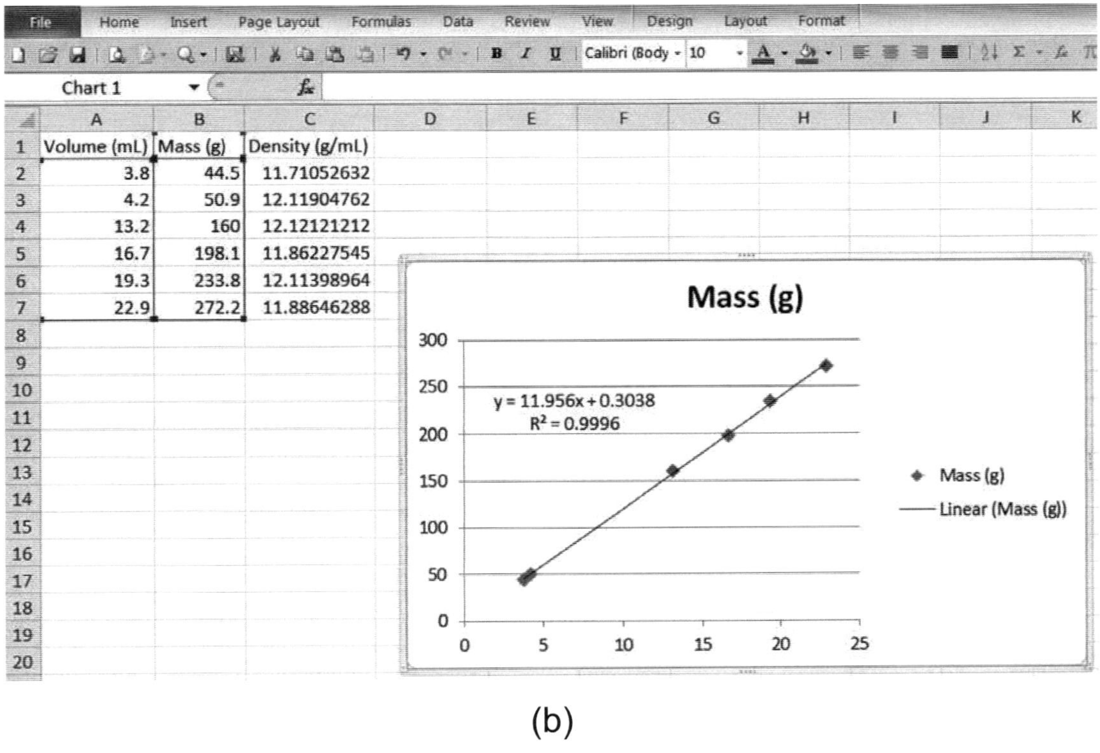

(b)

Figure 2.   Make a graph and add the trendline.

4. Use of Analytical Balances

One of the most common and important operations encountered in the chemical laboratory is that of determining the weight (mass) of given objects or samples. For that purpose, a suitable balance must be used. In this experiment three types of balances will be employed: the analytical balance, the triple beam balance, and the top–loading balance. In many of the experimental procedures in this course, accurate quantitative values are required. In such instances the analytical balance will be employed in weight determinations.

In this laboratory single pan analytical balances which weigh to the nearest 0.0001 g (0.1 mg) will be used in many cases. These balances are expensive and delicate (Figure 3). Your laboratory instructor will demonstrate the proper technique for using the specific balance model in your laboratory. The following instructions are given for general care of the analytical balance.

1) Do not overload the balance. Most analytical balances are designed for a maximum load of 100 or 120 g.
2) Always use a piece of weighing paper on the balance pan. Never weigh any chemical or moist object directly on the balance pan. Corrosive liquids and solids must be placed in a vapor–tight, pre–weighed container before weighing.

Figure 3. Analytical balance.

3) Close the balance doors before reading the balance weight. Air currents must be avoided.

4) Do not lean on or jar the balance table. Either will affect the weighings made on your balance and those of your neighbor.

5) Allow the sample to warm up or cool down to room temperature before weighing (temperature changes affects reading on an analytical balance).

6) Use the same balance for all of your weighing for a particular experiment.

7) Keep the balance clean at all time. Immediately clean up any chemical spilled over the balance. After you have finished weighing, check the following:

a) No object is left on the weighing pan;

b) The balance pan is clean;

c) Tare the balance.

Weighing by Difference

Weighing by difference is particularly useful for weighing samples that are hygroscopic or unstable when exposed in the air. To weigh by difference, the mass of a weighing bottle containing solid reagent is measured (m_1). After some of the sample is transferred to another container, the mass of the weighing bottle and the remaining solid is measured (m_2). The difference in the two masses ($m_1 - m_2$) represents the mass of solid reagent transferred to the vessel. Data should be recorded in your lab notebook as follows:

Initial mass of weighing bottle and sample 23.4567 g;

Mass of weighing bottle and sample after transfer 23.2898 g;

Mass of sample transferred $23.4567 - 23.2898 = 0.1669$ g.

5. Use of Triple Beam Balances

The primary features of the triple beam balance are the object platform, three weight beams, and the indicating pointer and zero scale (Figure 4). The weight of an object is established by positioning weights along the beams until balance is achieved. The middle and rear beams have notches in which the movable weights must only be placed. The sliding weight of the front beam can be positioned anywhere along the graduated portion of the beam.

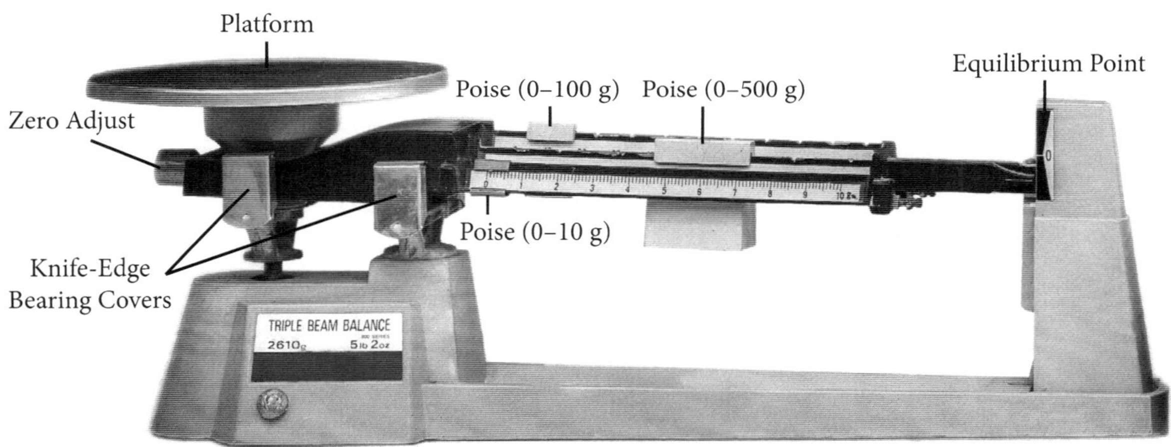

Figure 4. Triple beam balance.

Procedure to operate a triple beam balance:

1) Check the rest point with platform empty and all beam weights at zero. It is convenient, though not necessary to have the rest point at zero. Adjust the screw–type weight under the balance platform if necessary.

2) Place the object to be weighed on the platform.

3) Move the weight on the middle beam one notch at a time until the balance pointer deflects and stays below the zero mark, then bring the weight one notch back. Repeat the same procedure with the weight on the rear beam.

4) Adjust the sliding weight of the front beam so that pointer swings equally above and below the zero point. If the balance was not zeroed initially, pointer must return to initial rest point.

5) Record the weight as the sum of the weights on the three beams.

6) Return the weights to their original positions.

6. Use of Top–loading Balances

This type of balance is one of the easiest and quickest to use (Figure 5). In the latest models of top–loading balances, the operation procedure involves simply placing the object on the platform and reading the weight directly on the optical display. The balance must be level and in some cases it must be zeroed before weighing.

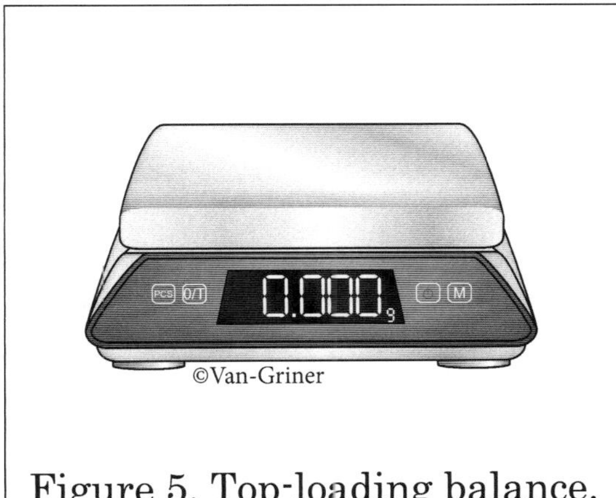

Figure 5. Top-loading balance.

7. How to Read Graduated Cylinders

To assist in reading a graduated cylinder or a buret, a piece of white paper can be placed behind it. Read the bottom of the meniscus. To make an accurate reading, your eye should be level with the bottom of the meniscus

(Figure 6). If the solution is very dark and it is difficult to determine the bottom of the meniscus, it is OK to read the top of the meniscus. Whatever part of the meniscus you read, be consistent.

To read a volume correctly, you must **estimate the volume to the nearest 1/10 of the smallest scale division**. For example, the volume from the graduated cylinder in Figure 6 is between 30.7 and 30.8 mL. The smallest scale division is 0.1 mL. So the volume reading should be 30.73 mL with the last digit "3" being an estimate. Different people estimate the same volume a little differently, so 30.74 mL or 30.72 mL are acceptable readings as well.

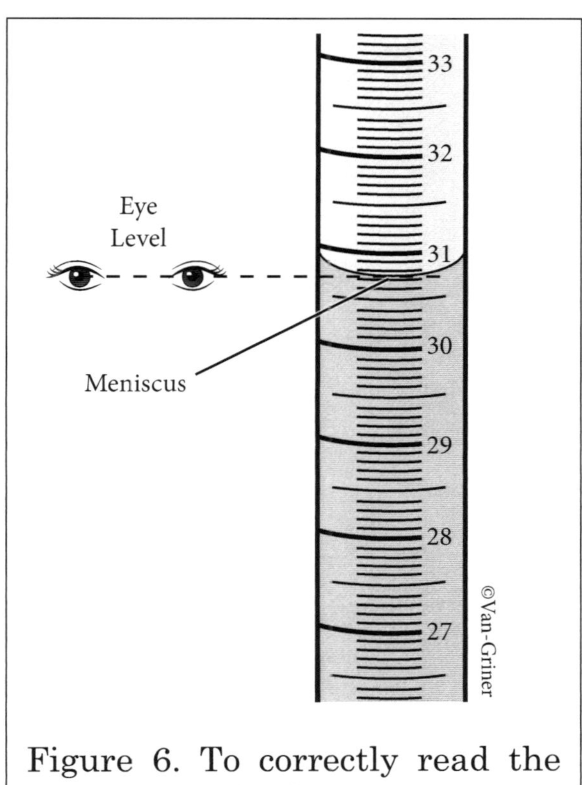

Figure 6. To correctly read the graduated cylinder.

These concepts and basic techniques will be repeatedly used in all the Chem 121L experiments, and you will experience more advanced techniques from them. Work safely and enjoy chemistry!

Experiment 1 Separation and Purification

Discussion

The four major steps involved in the analysis of a substance are (1) collection of the sample, (2) preparation for measurement, (3) measurement, and (4) evaluation of data. This experiment is designed to illustrate four techniques commonly used to purify and otherwise prepare a sample for measurement. These are decantation, filtration, coagulation, and distillation. A brief discussion of these techniques along with two other commonly used methods, extraction and sublimation, is presented below. The other steps in analysis: collection, measurement, and evaluation of data are subjects of future experiments and discussions.

Decantation

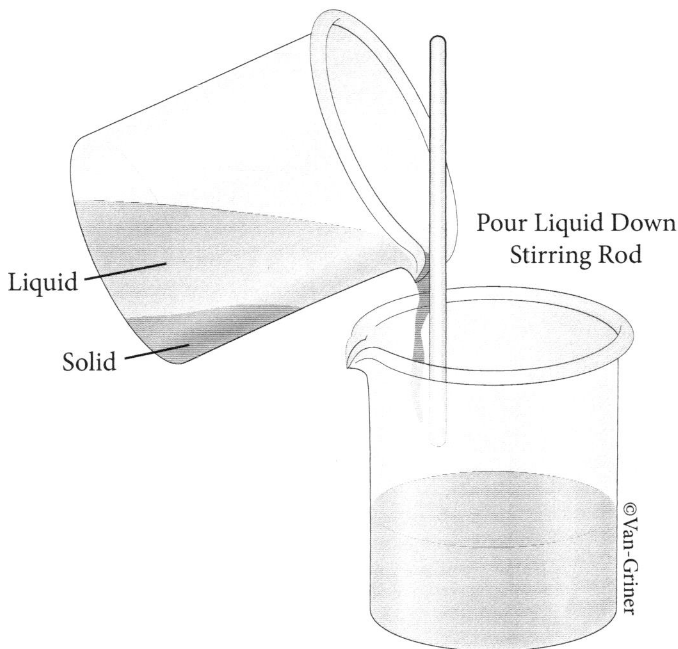

Figure 7. Decantation.

Many liquids can be easily separated from unmixed solids by decantation, which is by carefully pouring the liquid off and leaving the solids behind. For this method to be effective, the solids must first be allowed to settle momentarily. An example would be a mixture of sand and water. The decantation process is illustrated in Figure 7.

Filtration

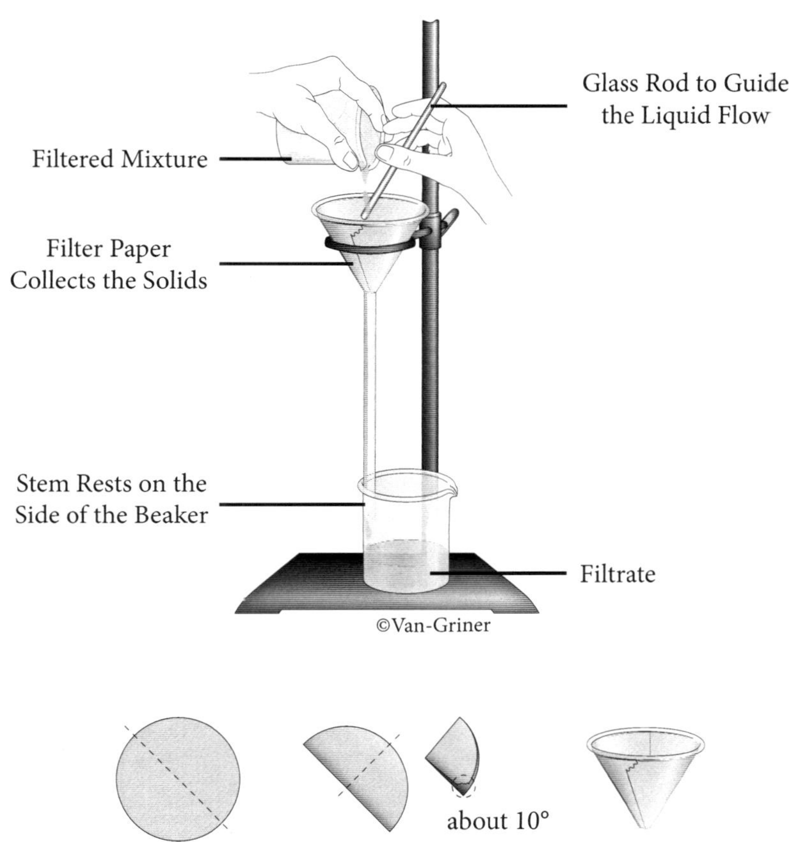

Figure 8. Gravity filtration (top) and folding the filter paper (bottom).

A solution quite often contains solid matter whose density is almost the same as the density of the liquid portion. When this is the case, the solids

do not settle to the bottom, and separation by decantation is impossible. Under this circumstance filtration is useful. Filters are composed of insoluble solids whose pore size allows the liquid portion to pass through but not the solid portion. Quite often it is the solid portion that is desired for analysis. In either event a separation is effected. A typical apparatus for gravity filtration is shown in Figure 8.

Coagulation

A solution may contain suspended particles of colloidal size that pass through filter paper, thus rendering filtration ineffective as a separation technique. These colloidal particles are kept suspended by the kinetic energy of the solvent molecules and exhibit "Brownian" movement. However, they can be forced to settle by the formation of a gelatinous precipitate in the solution, e.g., $Al(OH)_3$, which as it falls to the bottom of the container carries down most of the insoluble matter. This process is known as coagulation and is used in many municipal water treatment plants as one of several purification steps.

Distillation

Distillation is a common purification technique that is used in the separation of volatile liquids from undesirable impurities that are usually nonvolatile. A typical experimental set–up for a simple distillation is shown in Figure 9. The volatile liquid to be purified is vaporized in the round bottom flask, converted back to a liquid in the condenser, and finally collected in the receiving flask. The nonvolatile impurity remains in the original mixture. To avoid bumping, especially when solid material is

present in the bottom of the distillation flask, 2 or 3 small **boiling chips** are added to the flask before heating.

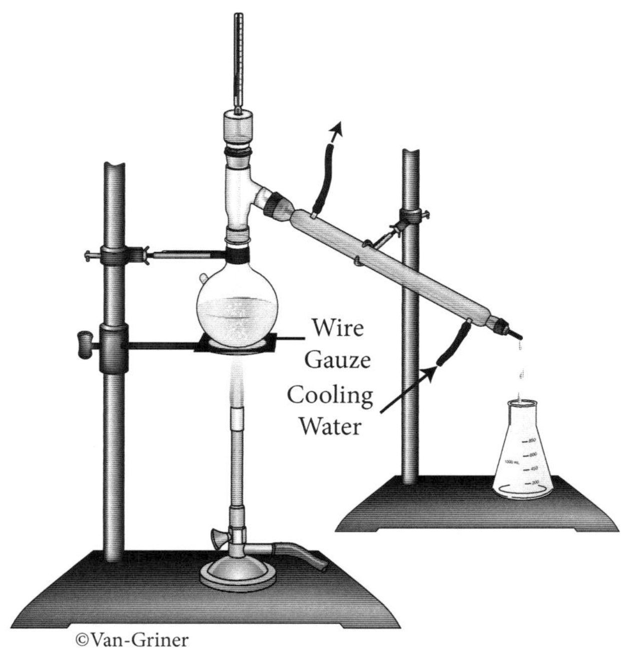

Figure 9. Apparatus for a simple distillation.

Extraction

Often it is necessary to remove a component of a solution that is not removable by the above methods. Consider a solution consisting of iodine, I_2, dissolved in H_2O. To remove I_2, an immiscible solvent such as carbon tetrachloride CCl_4, is added to the I_2/H_2O solution and mixed in a separatory funnel (Figure 10). The I_2 is much more soluble in CCl_4 than in water. Two liquid layers appear, water and CCl_4, with I_2 now extracted largely into the non−aqueous layer. Carbon tetrachloride has a much higher density than water, so it settles to the bottom while the water layer floats on top. A separation has been effected. Your instructor will demonstrate this technique.

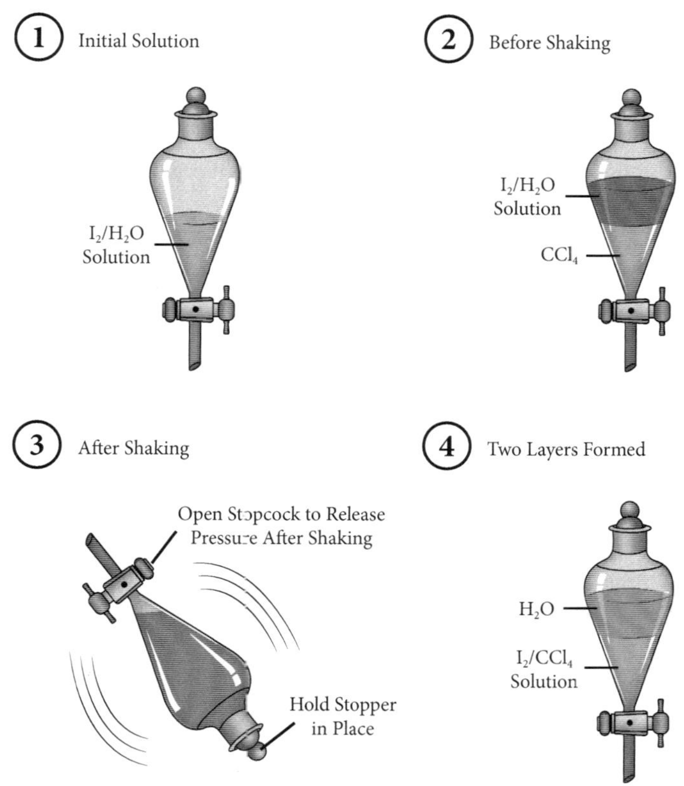

Figure 10. Extraction of iodine in water (follow the procedures of ① to ④).

Sublimation

Sublimation is sometimes a convenient method for separating two solids from one another. Sublimation is the process whereby molecules pass directly from the solid to the gaseous state. A mixture of sodium chloride and I_2 is placed in a beaker. A watch glass is placed over the beaker and a piece of ice placed on top of the watch glass. The beaker is heated. Pure iodine crystals will form on the bottom of the watch glass. The basis for this separation is that iodine can be converted directly from the solid to the gaseous state (and back again), but the sodium chloride cannot.

Sublimation is exhibited by other substances, the most common being "dry ice" or solid CO_2.

Procedure

1. Decantation. A solution to be purified is provided in the laboratory. It is a mixture of sand, dirt, salt, and water. In a beaker, obtain approximately 100 mL of this solution. Allow it to remain undisturbed for several minutes. Carefully pour the liquid portion into another beaker without disturbing the insoluble sediment. Discard the sediment into the crocks provided. Keep the decanted solution for use in the next portion of the experiment.

2. Filtration. Properly fold a sheet of filter paper (See Figure 8) and place it in a funnel. Pour the solution obtained in part 1 through the filter paper being careful not to allow it to overflow. Collect the solution into a 100 mL graduated cylinder.

3. Coagulation. From the reagents provided mix 8 mL of saturated $Ca(OH)_2$ with 2 mL of 0.1 M $Al_2(SO_4)_3$ in a test tube and shake. Add this solution to the solution from part 2 and allow it to settle. The reaction occurring is represented by the equation:

$$Al_2(SO_4)_3 \text{ (aq)} + 3 \; Ca(OH)_2 \text{ (aq)} = 2 \; Al(OH)_3 \text{ (s)} + 3 \; CaSO_4 \text{ (aq)}$$

4. Decantation and Precipitation. Decant approximately 50 mL of clear solution from part 3 and add several drops of $AgNO_3$, sufficient to form a white precipitate. The reaction occurring is:

$$Ag^+ \text{ (aq)} + Cl^- \text{ (aq)} = AgCl \text{ (s)}$$

The Cl⁻ ions are present because the original solution contained some dissolved NaCl.

5. Distillation. Place the solution from part 4 in the distillation flask and assemble your distillation apparatus. Your instructor will demonstrate the use of your gas burner, then proceed to distill the solution until approximately 10 mL of distillate has been collected. Remove the burner and test the distillate (the liquid that distilled) with $AgNO_3$ for the presence of dissolved chloride ions (Cl^-).

Gas Burner

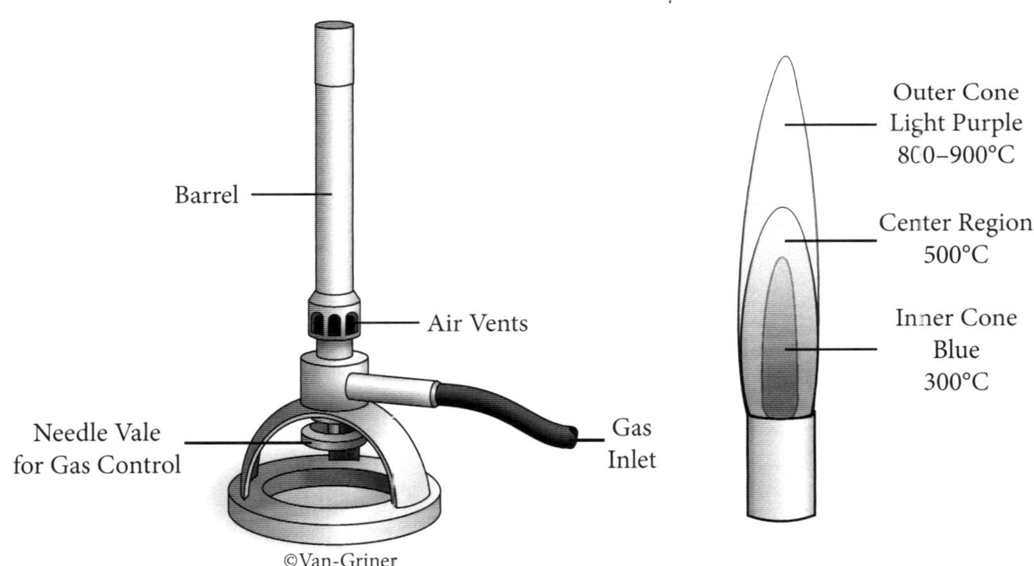

Figure 11. Gas burner.

A gas burner similar to the one depicted above (Figure 11) will be used frequently in this course. In this burner, the gas flows through the gas inlet into the barrel and is controlled by the gas needle valve. The flow of air into the barrel is controlled by turning the barrel itself which adjusts the air

inlets. Proper control of the gas needle valve and air inlets will produce the desired combustion mixture. The following procedure should be followed in lighting the burner.

To light the burner:
1. Connect the gas pipe and the burner with a good rubber tube;
2. Stay away from the burner;
3. Slightly open the gas needle valve (about one half turn) on the burner; put a lighted match from the side (slightly above the top of the barrel), then open the gas valve slowly. You should see a flame above the top of the barrel.
4. Gradually open the air inlets until the flame takes on a blue color with a well–defined blue cone. The top of this cone is the hottest point of the flame.
5. If the supply of air or gas is too large, the flame will go out usually with a noise. If this occurs, turn off the gas immediately and start over. When relighting, be sure to shut off the air inlets initially.

Experiment 2   Isotopes and Atomic Mass

In this laboratory you will attempt to determine the relative abundance of a fictional element and calculate its atomic mass. This laboratory will model an element with three isotopes. The "isotopes" to be counted and weighed in this experiment are pinto beans, red beans, and split peas. We will experience **three methods** to determine the atomic mass. Table 1 is for **Method 1**, Tables 2 – 4 are for **Method 2**, and Table 5 is for **Method 3**.

Procedure

1. Obtain a sample of the three–isotope element "legumium" from the instructor.
2. Make four (4) measurements of the atomic mass: each should be a **random** fraction of the total sample. Measure the mass and count the number of isotopes in each sample, and then calculate the average mass (atomic mass). The top–loading balance can be used in this experiment. Enter the data in Table 1 below. This finishes the **Method 1** of finding the atomic mass through fractions of the legumium isotope mixture samples.

Table 1 Find the average atomic mass through fractions of isotope samples

Sample #	Measured mass (g)	Number of atoms	Atomic mass (g)
1			
2			
3			
4			
Average atomic mass $\longrightarrow$			

3. Now we work on **Method 2** to determine the atomic mass of the sample from the relative abundance of each isotope and the mass of each isotope. To do this, completely separate the whole bottle isotopes: pinto beans, red beans, and split peas. Determine the average mass of a particle of each isotope (average isotope mass). Enter your data into Table 2.

Table 2 The average mass of each isotope

Isotope	Total mass (g)	Number of isotopes	Average isotope mass (g)
Pinto bean			
Red bean			
Split pea			

4. Determine the relative abundance of each isotope in the sample by dividing the number of particles of each isotope by the total number of particles and then multiplying by 100. This will give you the percentage of each isotope in legumium. Enter your data in Table 3.

Table 3 Find the relative abundance of the isotopes in the sample

Isotope	Number of isotopes	Relative abundance (%)
Pinto bean		
Red bean		
Split pea		
Total		100.0

5. Determine the atomic mass of the legumium by multiplying the relative abundance of each isotope by the average mass of each isotope and then adding together the contributions of each isotope. Enter your data in Table 4. This finishes the **Method 2.**

Table 4 Determine the atomic mass by the contributions of isotopes

Isotope	Average isotope mass (g) (from Table 2)	Relative abundance (%) (from Table 3)	Contribution (g)
Pinto bean			
Red bean			
Split pea			
Atomic mass (g) $\longrightarrow$			

6. The atomic mass of legumium can also be experimentally determined by finding the mass of your entire sample and dividing by the total number of isotopes in the sample. The calculated value is the "true" atomic mass of legumium. This is the **Method 3**.

Table 5 Find the atomic mass of by counting the entire sample

Entire sample	Measured mass (g)	Number of atoms	Atomic mass (g)

Lab report

In addition to the required contents, the following information should also be included in your lab report:

1. The true value for the atomic mass of legumium is obtained from Step 6 (Table 5). The value is ___________.
2. Calculate the average atomic mass from Step 2 (Table 1). The average atomic mass is ___________.
3. Determine the error in the average mass calculated in Step 2 (Table 1). To do this you must first calculate the deviations in each of the measurements.

$$\text{Deviation} = \left| \text{experiment value} - \text{average value} \right|$$

Deviation #1   = ___________ ;

Deviation #2   = ___________ ;

Deviation #3   = ___________ ;

Deviation #4   = ___________ ;

The average of the four deviations is then calculated.

Average deviation = ___________ .

The measured result is then reported as:

Atomic mass = Average atomic mass $\pm$ Average deviation

Atomic mass = ___________ $\pm$ ___________ .

4. Does your calculation of the atomic mass in Step 2 (method 1) agree with the true atomic mass (determined in Step 6)? Are these two values within the average deviation? If not, explain it.

5. Does your calculation of the atomic mass in Step 5 (Table 4) agree with the true atomic mass you determined in Step 6 (at least within the error)? If not, explain it.

Experiment 3 Determination of the Density of Solid Subjects

Discussion

The density of any substance is defined as the mass of the substance per unit volume. In other words, density is the name given to the property which is the ratio of the mass to volume. This can be written as: $d = m/V$, where d is the density, m is the mass, and V is the volume. The units used to express density depend on the units used in measuring the mass and the volume. In the laboratory, mass is usually measured in grams (g) and volume in cubic centimeters (cm^3) or milliliters (mL). Therefore, the most common units for density are g/cm^3 or g/mL. Since 1 cm^3 is equal to 1 mL, these two units for density are the same.

Both the mass and the volume depend on the size of the sample. The density, however, is independent of the size of the sample since it is the ratio between mass and volume. Consider two samples of gold. One has a volume of 4.00 mL and a mass of 77.2 g. The other has a volume of 1.00 mL and a mass of 19.3 g. The density of each sample can be calculated as

$d_1 = m_1/V_1 = 77.2$ g/4.00 mL $= 19.3$ g/mL, and

$d_2 = m_2/V_2 = 19.3$ g/1.00 mL $= 19.3$ g/mL.

The two samples of gold have different masses and volumes, but the density is the same for each sample. As a result, density can be used to distinguish one pure substance from another. In this experiment you will determine the densities of zinc and aluminum cylinders. The masses will be measured using the analytical balance (for zinc) and the top–loading balance (for

aluminum). The volume of solids can be determined by one of two methods, and both of these methods will be employed in this experiment.

a. using the geometrical formula for the volume of regularly shaped solids after measuring the object's dimensions.

b. the water displacement method, in which the solid is submerged into a premeasured volume of water and the new volume is then measured. The increase in the volume is equal to the volume of the object. This method is very convenient for solids with irregular shapes.

Whenever a straight line can be drawn through a set of data points, the data can be described in terms of the mathematical equation $Y = a X + b$. In this expression Y represents the values plotted on the ordinate (vertical axis), a is the slope of the straight line, and b is the intercept of the straight line at the ordinate, i.e. the value of Y when X is zero. The relationship between the two properties mass and volume for various samples of the same substance can thus be determined by a graphical approach. If the mass and the volume of each sample are determined, the mass of each sample can be plotted on the ordinate and the corresponding volume on the abscissa. The density of the substance can then be determined from the graph as the slope of the curve. Remember that the slope of a straight line is calculated as the difference in two Y values divided by the difference in the corresponding two X values.

$$\text{Slope } a = \Delta y / \Delta x = (y_2 - y_1)/(x_2 - x_1) = \text{density}$$

Procedure

Obtain a set of zinc and aluminum cylinders (5 of each).

1. Density of zinc

In the determination of the density of zinc you will measure the mass using the analytical balance and determine the volume by measuring the length and diameter of the cylinders.

1) Weigh the zinc cylinders individually on the analytical balance to the nearest 0.0001 g, and record the values in your notebook. Note: there are two kinds of analytical balances in the lab. One has capacity of 120 g, and another is 220 g. The longest zinc cylinder may weigh more than 120 g; in that case, you can find the analytical balances with capacity of 220 g to measure the mass.

2) With a Vernier caliper (Figure 12), measure the diameter of the cylinders to the nearest 0.005 cm (they should all have the same diameter) and record this value in your notebook.

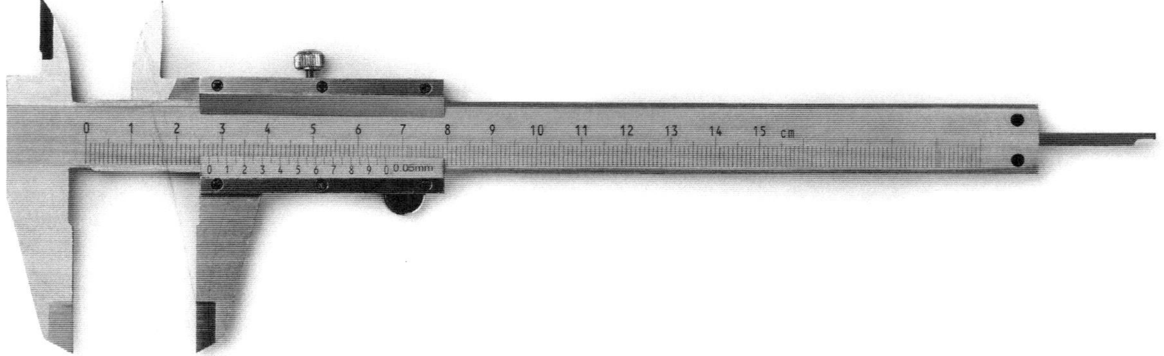

Figure 12. Use of a Vernier caliper.

3) Again using the Vernier caliper, measure the length of each cylinder to the nearest 0.005 cm, and record the values in your notebook.

2. Density of aluminum

In the determination of the density of aluminum you will measure the mass with the top–loading balance and determine the volume by the displacement method.

1) Weigh the aluminum cylinders individually on the top–loading balance to the nearest 0.01 g, and record the values in your notebook.

2) Place enough water that can cover the longest aluminum cylinder in a 100 mL graduated cylinder. Record the volume of the water to the nearest 0.1 mL by reading the bottom of the meniscus (see Figure 6 on page 12). **GENTLY** place the aluminum cylinder into the graduated cylinder. **You will have to tip the graduated cylinder on an angle and slowly slide the aluminum cylinder down into water in order not to break the graduated cylinder.** Read the new volume and determine the volume of each piece of aluminum metal. Record each of the values in your notebook.

Calculation

1. Density of zinc

Calculate the volume of each cylinder (in cm^3) by using the formula

$V = (\pi \times d^2 \times h)/4$, where d is the diameter of the cylinder, h is the height (length) of the cylinder and $\pi = 3.1416$. Record these volume values in your notebook.

Calculate the density of each cylinder by dividing the measured mass by the calculated volume, and record the density values in your notebook.

Graph the mass of the zinc cylinder as the ordinate versus the volume of the zinc cylinder as the abscissa (refer to page 10 for "Graphical Analysis by Excel Spreadsheets"). You will have to choose appropriate scales in order to construct a graph of reasonable size. Use the linear regression to fit all the data points. Determine the slope for the density of the zinc from the graph.

2. Density of aluminum

Since the mass and volume have already been measured, determine the density of each cylinder by dividing the mass of the cylinder by the volume of the cylinder. Write down these density values in your notebook. You will find it very helpful to graph both the zinc and aluminum values of the same graph and compare the densities (slopes) of the two metals.

Lab report

In addition to the required contents, the following information should also be included in your lab report:

1. Density of zinc

Diameter of zinc cylinders (cm)   _______________

Cylinder #	Mass (g)	Length (cm)	Volume (cm^3)	Density (g/cm^3)
1				
2				
3				
4				
5				

Average density (g/cm^3)   _______________

Density of zinc (from graph) (g/cm^3)   _______________

2. Density of aluminum

Cylinder #	Mass (g)	Volume (mL)	Density (g/cm^3)
1			
2			
3			
4			
5			

Average density (g/cm^3) ______________

Density of aluminum (from graph) (g/cm^3) ______________

Experiment 4   Specific Gravity and Density of a Solution

Discussion

In the previous experiment, the density of solid zinc and aluminum were determined. The density of liquids can also be readily determined experimentally. However, like solids, the determination of the volume of the substance is the most difficult measurement. Since density is defined as mass per unit volume, $d = m/V$, it has the units of grams per milliliter (g/mL).

Specific gravity is defined as the ratio of the mass of a substance to the mass of an equal volume of a reference substance (usually water) when the temperature is constant. Since specific gravity is a ratio of masses, specific gravity is a dimensionless quantity. A specific gravity bottle (also called pycnometer, see Figure 13) can be conveniently used to determine the specific gravity of an unknown solution. The weight of the empty bottle is first determined. Then the bottle is filled successively with water followed by the unknown solution and the weight of each substance occupying the bottle is determined.

Figure 13.
The specific gravity bottle.

$$\text{Specific gravity} = m_{solution}/m_{water} \text{ (with equal volume)}$$

We know d = m/V, then m = d × V. Therefore when the volumes of the two solutions are equal the specific gravity of the solution is the ratio of the density of the solution to the density of water (at the same temperature), i.e.

$$\text{Specific gravity} = d_{solution}/d_{water}$$

At room temperature the density of water can be regarded as 1 (check page 41 for the accurate values at different temperatures); therefore, the specific gravity and density of the solution have the same numerical value. At temperatures far from room temperature, the specific gravity and density are not equal.

Procedure

In this experiment you will determine (1) the specific gravity of an unknown solution, and (2) the density of that same unknown solution.

1. Obtain a specific gravity bottle (pycnometer) from the instructor.
2. Ascertain the bottle and the plug are clean and dry; wipe the outside with a towel before weighing. Weigh the bottle three times on an analytical balance, wiping the outside before each weighing, and calculate the mean weight of the empty bottle.
3. Completely fill the bottle with distilled water, insert plug, blot any excess water and weigh. Empty the water and repeat this step two additional times.
4. Rinse the bottle and plug thoroughly with the unknown solution three times and discard the wastes. This can be done by filling the bottle halfway, inserting the plug, and shaking. Then fill the bottle with the

unknown solution, insert the plug, wipe off excess solution and weigh. Empty the unknown solution and refill and weigh for two additional times.

5. Rinse the bottle and plug thoroughly with distilled water three times and put them back into the small beaker.

Lab report

In addition to the required contents, the following information should also be included in your lab report:

1. Determination of specific gravity of unknown solution

Measurement	1	2	3
Weight of bottle (g)			
Mean weight of bottle (g)			
Weight of bottle + water (g)			
Mean weight of bottle + water (g)			
Weight of water (g)			
Weight of bottle + unknown (g)			
Mean weight of bottle + unknown (g)			
Weight of unknown (g)			
Specific gravity of unknown			

2. Determination of density of unknown solution

Weight of water (from above) (g)	
Temperature of water (°C)	
Density of water at recorded temperature (see next page) (g/mL)	
Volume of bottle (mL) *	
Weight of unknown (from above) (g)	
Density of unknown (g/mL)	

* Do not use the nominal volume marked on the specific gravity bottle; calculate it from the mass and the density of pure water.

Density of Water (g/mL) vs. Temperature (°C)

To find the density of water at 25.4°C, you first find the whole degree by searching down the left hand column until you reach "25". Then you slide across that row until you reach the column labeled "0.4". The density of water at 25.4°C is 0.996941 g/mL. (*Handbook of Chemistry and Physics*, 53rd Edition, p. F4)

	0.0	0.1	0.2	0.3	0.4	0.5	0.6	0.7	0.8	0.9
10	0.999700	0.999691	0.999682	0.999673	0.999664	0.999654	0.999645	0.999635	0.999625	0.999615
11	0.999605	0.999595	0.999585	0.999574	0.999564	0.999553	0.999542	0.999531	0.999520	0.999509
12	0.999498	0.999486	0.999475	0.999463	0.999451	0.999439	0.999427	0.999415	0.999402	0.999390
13	0.999377	0.999364	0.999352	0.999339	0.999326	0.999312	0.999299	0.999285	0.999272	0.999258
14	0.999244	0.999230	0.999216	0.999202	0.999188	0.999173	0.999159	0.999144	0.999129	0.999114
15	0.999099	0.999084	0.999069	0.999054	0.999038	0.999023	0.999007	0.998991	0.998975	0.998959
16	0.998943	0.998926	0.998910	0.998893	0.998877	0.998860	0.998843	0.998826	0.998809	0.998792
17	0.998774	0.998757	0.998739	0.998722	0.998704	0.998686	0.998668	0.998650	0.998632	0.998613
18	0.998595	0.998576	0.998558	0.998539	0.998520	0.998501	0.998482	0.998463	0.998444	0.998424
19	0.998405	0.998385	0.998365	0.998345	0.998325	0.998305	0.998285	0.998265	0.998244	0.998224
20	0.998203	0.998183	0.998162	0.998141	0.998120	0.998099	0.998078	0.998056	0.998035	0.998013
21	0.997992	0.997970	0.997948	0.997926	0.997904	0.997882	0.997860	0.997837	0.997815	0.997792
22	0.997770	0.997747	0.997724	0.997701	0.997678	0.997655	0.997632	0.997608	0.997585	0.997561
23	0.997538	0.997514	0.997490	0.997466	0.997442	0.997418	0.997394	0.997369	0.997345	0.997320
24	0.997296	0.997271	0.997246	0.997221	0.997196	0.997171	0.997146	0.997120	0.997095	0.997069
25	0.997044	0.997018	0.996992	0.996967	0.996941	0.996914	0.996888	0.996862	0.996836	0.996809
26	0.996783	0.996756	0.996729	0.996703	0.996676	0.996649	0.996621	0.996594	0.996567	0.996540
27	0.996512	0.996485	0.996457	0.996429	0.996401	0.996373	0.996345	0.996317	0.996289	0.996261
28	0.996232	0.996204	0.996175	0.996147	0.996118	0.996089	0.996060	0.996031	0.996002	0.995973
29	0.995944	0.995914	0.995885	0.995855	0.995826	0.995796	0.995766	0.995736	0.995706	0.995676
30	0.995646	0.995616	0.995586	0.995555	0.995525	0.995494	0.995464	0.995433	0.995402	0.995371

Experiment 5   Law of Definite Proportions

Discussion

Dalton's atomic theory includes the concepts that atoms of a given substance all have the same weight and that compounds are formed by combinations of atoms. The experimental information referred to as the Law of Definite Proportions supports these two concepts. Without exception, careful study of binary compounds bonded by conventional covalent or ionic bonding has shown that each compound always contains the same weight percentages of the two elements entering into chemical combinations to form the compound. The purpose of this experiment is to repeatedly synthesize a binary compound and to determine its empirical formula.

You will chemically combine two elements, copper and iodine. Copper is a soft metal that will be used in the form of a fine wire. Iodine is a solid that sublimes readily, that is, with gentle heating the solid converts to vapor. In this experiment the iodine vapor will be brought into contact with the copper wire. A new compound is formed in a chemical reaction, and it appears in the form of a **white solid** that adheres to the surface of the copper wire. The copper wire will now weigh more than it did originally because of the adhering compound. Note that the additional weight is due only to the iodine atoms in the compound since the copper atoms in the compound were originally part of the weight of the copper wire. The copper atoms have simply changed from an association with other copper atoms in the metal to an association with iodine atoms in the compound.

The chemical properties of atoms of copper as they are found in the compound are different from the chemical properties of those found in the metal. In particular, the atoms of copper as they are found in the compound will react with a solution of sodium thiosulfate ($Na_2S_2O_3$) to form a new water soluble substance containing the copper atoms (copper wire will not react with sodium thiosulfate solution). As a result the copper compound dissolves while the unreacted copper does not. Now, the copper wire stripped of its coating of the compound will weigh less than the original copper wire because some copper atoms have been removed from the surface. In fact, the difference in weight is just the weight of copper atoms which entered into chemical combination with iodine atoms to form the white solid compound. Thus, the weight the copper wire gains upon being coated with the compound is due to the iodine atoms in the compound, and the weight the copper wire has lost when the compound is stripped off is due to the copper atoms in the compound.

According to the Law of Definite Proportions, whatever weight of copper and iodine may combine to form the compound, the ratio of the weight of iodine to the weight of copper will be the same in all cases. To test this experimentally, carry out the following procedure at least three times and examine the data in terms of the weight of iodine that reacts with one gram of copper in each synthesis of the compound.

Procedure

1. Take a piece of #24 copper wire about 1 meter in length. Wind the wire tightly around a pen or pencil so as to form a close spiral of wire. Leave about 5 cm of straight wire to be used as a handle. Form a short bend in

the end of the handle so that the coil of wire can be suspended from the top rim of a test tube.

2. Dip the copper coil in a dilute solution of nitric acid (2 M) for around fifteen seconds. Rinse the coil by dipping it in distilled water. Next dip the coil in acetone and let the coil air–dry. Weigh the copper coil on the analytical balance. Weigh the coil to the nearest $\pm$ 0.1 mg.

3. In the laboratory hood you will find test tubes (25×150 mm) containing iodine being warmed so that iodine vapors (caution: iodine solid and vapor are toxic) rise near the top of the test tube. Adjust the length of the handle and hook on the coil of copper wire so that the bottom of the coil is at least 3 cm above the bottom of the test tube. Insert the coil in the test tube and suspend it by hooking it on the edge of the test tube. Leave the coil in contact with the iodine vapor for at least two minutes. Carefully remove the wire taking care not to shake or jar it so severely that some of the coating may be lost. To remove the excess iodine crystals, if necessary, hold the coated wire in the warm air from the heat gun, avoiding excess vibration of the coil. Weigh the coated copper wire on the analytical balance to the nearest 0.1 mg.

4. Immerse the coated copper coil in a 0.5 M sodium thiosulfate solution for about three minutes, with occasional swirling. Be sure that all of the coating is covered by the thiosulfate solution. After the bright surface of copper reappears, remove coil from the solution and rinse it in distilled water. Then dip the coil in acetone and air–dry. Weigh the copper coil on the analytical balance to the nearest 0.1 mg.

5. In your notebook, calculate the relative weight of iodine to copper to ascertain if your techniques are correct.

6. Repeat steps 3 and 4 two or more times. Calculate the relative weight of iodine to copper in each of the experiments. In order to obtain reasonable values, mass ratios of reacted iodine and copper should range between 1.9 and 2.1 for iodine to copper.

7. Determine the formula for the compound formed.

Lab report

In addition to the required contents, the following information should also be included in your lab report:

Measurement	Trial 1	Trial 2	Trial 3
Mass of Cu wire (g)			
Mass of Cu wire plus compound (g)			
Mass of Cu wire with compound removed (g)			
Mass of I in compound (g) – A			
Mass of Cu in compound (g) – B			
Reacted mass ratio of I/Cu – A/B			
Moles of I in compound (mol)			
Moles of Cu in compound (mol)			
Moles of I / Moles of Cu			
Mean value, Moles of I / Moles of Cu			
Deviation			
Mean deviation			
Empirical formula of the compound, CuI_n	$CuI\_\_$.		

Experiment 6   A Series of Chemical Reactions Involving Copper

Discussion

This experiment consists of a series of chemical reactions that provide an interesting sequential path from metallic copper through some of its common compounds and returning to copper again.

The following six equations describe the reactions the copper undergoes:

1. $Cu\ (s) + 4\ HNO_3\ (aq) = Cu(NO_3)_2\ (aq) + 2\ H_2O\ (l) + 2\ NO_2\ (g)$
2. $Cu(NO_3)_2\ (aq) + 2\ NaOH\ (aq) = Cu(OH)_2\ (s) + 2\ NaNO_3\ (aq)$
3. $Cu(OH)_2\ (s) = CuO\ (s) + H_2O\ (l)$
4. $CuO\ (s) + H_2SO_4\ (aq) = CuSO_4\ (aq) + H_2O\ (l)$
5. $CuSO_4\ (aq) + Zn\ (s) = Cu\ (s) + ZnSO_4\ (aq)$
6. $Zn\ (s) + H_2SO_4\ (aq) = ZnSO_4\ (aq) + H_2\ (g)$

Equation 1 describes the dissolution of solid copper in concentrated nitric acid to form a liquid solution. Although the substances are written as molecules and the equation as a molecular equation, the actual species present in solution are the ions of the compounds. Precipitation of the copper ions as insoluble $Cu(OH)_2$ occurs in Equation 2. Formation of insoluble precipitates usually tends to drive reactions to completion or to the right as described by the reaction.

Conversion of the $Cu(OH)_2$ solid, after filtering, to a different compound of copper, CuO, occurs in Equation 3 by heating to drive off water. Dissolution of the CuO with sulfuric acid is described in Equation 4. Finally the

conversion of the copper to its metallic form occurs in equation 5. Equation 6 is provided to show the production of gaseous hydrogen as often observed (due to the excess of H_2SO_4 used).

Procedure

1. $Cu(s) + 4\ HNO_3\ (aq) = Cu(NO_3)_2\ (aq) + 2\ H_2O\ (l) + 2\ NO_2\ (g)$

   Obtain a sample of copper wire (about 0.4 to 0.6 g). Place it in a 400 mL beaker. Add (**in the hood**) 5 mL concentrated nitric acid (**care in handling**). Cover the beaker immediately with a watch glass and allow it to stand until all the copper dissolves. Remove the watch glass and allow the brown gas to disappear before taking the beaker to your desk. Then add 15 mL of distilled water. **NOTE: brown toxic fumes of NO_2 are produced from HNO_3 in water, so avoid contact and breathing the fumes. This gas is extremely poisonous, so use caution.**

2. $Cu(NO_3)_2\ (aq) + 2\ NaOH\ (aq) = Cu(OH)_2\ (s) + 2\ NaNO_3\ (aq)$

   Add 10 mL of sodium hydroxide solution (6 M) **cautiously** with stirring. Test with litmus paper, and if the solution is not basic (red litmus turns blue), add more sodium hydroxide in small portions until the solution is basic to litmus.

3. $Cu(OH)_2\ (s) = CuO\ (s) + H_2O\ (l)$

   Add 50 mL of distilled water. Bring the solution to boiling and boil gently **with constant stirring** until the blue color disappears or darkens. Filter while hot and discard the colorless liquid (refer to page 21 for operation); wash the solid with more water and discard the wash water.

4. $CuO\ (s) + H_2SO_4\ (aq) = CuSO_4\ (aq) + H_2O\ (l)$

   Punch a small hole with a wire in the bottom of the filter paper (this operation is only for this experiment; never punch a hole on the filter

paper for a regular filtration) and pour 10 mL of dilute sulfuric acid (3 M) through the filter, collecting the solution in a flask or beaker. Pour the same solution back through the filter, collecting as before and repeat the process until all of the solid has dissolved. Wash the filter paper with small portions (5 mL) of water and combine all solutions.

5. $CuSO_4$ (aq) + Zn (s) = Cu (s) + $ZnSO_4$ (aq)

    Zn (s) + H_2SO_4 (aq) = $ZnSO_4$ (aq) + H_2 (g)

    Add a piece of metallic zinc to your solution from step 4 and stir with a glass rod. Observe the evolution of hydrogen gas on the solid, and the formation of metallic copper (color change). You may discard the liquid to see the copper color better.

6. In your notebook, write a brief statement, including equations, color change, gas and precipitates generation, about each step in the sequence of chemical reactions.

Lab report

In addition to the required contents, the following information should also be included in your lab report:

1. What precautions must be taken when copper metal is dissolved in concentrated nitric acid?

2. In acid solution, red litmus paper appears what color?

3. Why does an excess of sodium hydroxide (NaOH) need to be added in the reaction of $Cu(NO_3)_2$ to form $Cu(OH)_2$?

4. When H_2SO_4 is added to CuO, the $CuSO_4$ that forms dissolves in water. The color of the mixture changes from __________ to __________.

Experiment 7 Hydrates

Discussion

Water has a strong attraction for many compounds because of its polar character and electronic structure. Water associated with solid chemical compounds occurs in one of two ways: simply as adsorbed water on the surface of the crystals, or chemically bound in larger amounts in certain compounds (usually ionic salts). Thus, many salts contain a definite number of moles of water as an integral part of the crystalline structure. For example, copper sulfate crystallizes as $CuSO_4 \cdot 5H_2O$; magnesium sulfate crystallizes as $MgSO_4 \cdot 7H_2O$; and borax has the composition of $Na_2B_4O_7 \cdot 10H_2O$. The water incorporated in the crystal structure is called water of hydration and the salts are known as hydrates. In recent years the theory has developed that this water results from the bonding of the water molecule with the cation and the strong electrostatic attraction between the water molecule and the anion.

Hydrates decompose on heating by releasing water to produce an anhydrous salt and water vapor. As the water of hydration is lost from a hydrate, the compound may go through several color changes, which correspond to the colors of the various hydrates formed by the salt. Hydrates that are not very stable will lose water of hydration at room temperature, and have a significant vapor pressure. Such hydrates are said to effloresce (lose water of hydration) when exposed to air. A few ionic compounds will absorb so much water from the atmosphere that they will eventually dissolve in their water of hydration; such substances are said to be delinquescent.

Hydrates that are moderately stable at room temperature may effloresce or deliquesce on being exposed to air. If the water vapor pressure of the air is greater than the vapor pressure of the hydrate, deliquescence occurs; efflorescence occurs when the conditions are reversed. Some hydrates are so stable that heating causes decomposition of the salt, rather than dehydration. This is illustrated by hydrated aluminum chloride, $AlCl_3 \cdot 6H_2O$, which may decompose in different ways. The equations that follow indicate two ways that this salt may decompose depending on the conditions that prevail.

$$AlCl_3 \cdot 6H_2O = AlOCl + 2\ HCl + 5\ H_2O$$
$$2\ AlCl_3 \cdot 6H_2O = Al_2O_3 \cdot 3H_2O + 6\ HCl + 6\ H_2O$$

Procedure

A. Efflorescence and deliquescence

1. Several chemical compounds have been placed on watch glasses in your laboratory for observation. Accompanying each is its correct formula and name. After approximately one hour determine:

a. Which of the salts deliquesced;

b. Which of the salts effloresced;

c. Which salt(s) neither effloresced nor deliquesced.

2. Place a small amount of cobalt (II) chloride ($CoCl_2 \cdot 6H_2O$) in a small clean test tube and gently heat with a Bunsen burner flame until the color change appears to be complete. Allow the system to cool while being exposed to the atmosphere. If little or no change is observed, add 1 or 2

drops of water. Record your observations in your notebook. Write the equation for the reactions observed when heating the salt. Write the equation for the reaction observed when adding water. How could this salt be used as a moisture indicator? Place the product in a large beaker found in the hood for reuse in another laboratory.

B. Determination of the formula of a hydrate

Weigh a clean, dry evaporating dish on the analytical balance to the nearest 0.1 mg. Place 1 ~ 2 g of an unknown crystalline hydrate (to be supplied by instructor, write down the number of the unknown hydrate) in the weighed dish and reweigh to the nearest 0.1 mg. Place the dish on the wire triangle (or wire gauze) supported by a ring stand or tripod and cover with a watch glass (Figure 14). Heat the dish slowly for approximately 15 ~ 20 minutes using a small colorless flame, until no moisture

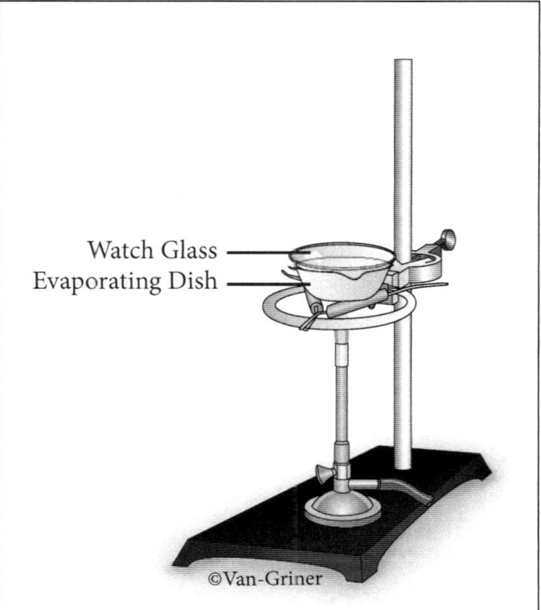

Figure 14. Experimental apparatus for the dehydration of a hydrate.

is visible on the watch glass. If the hydrate melts, heat slowly to prevent splattering of the salt. **Care must be taken to prevent the flame from touching the watch glass as the glass will break.**

Allow the solid to cool to room temperature and weigh immediately using an analytical balance. Reheat the dish for about 5 minutes. Do not overheat as the salt may decompose. Cool and reweigh. Repeat the heating, cooling, and weighing until two consecutive weighings agree within 1 mg (0.001 g).

Lab report

In addition to the required contents, the following information should also
be included in your lab report:

Mass of dish (g) __________________

Mass of dish + hydrate (g) __________________

Mass of dish + contents after heating (g) __________________

Mass of water driven off (g) __________________

Moles of water driven off (mol) __________________

Mass of anhydrous salt (g) __________________

Moles of anhydrous salt (mol) __________________

Formula weight of anhydrous salt

    (see the table below) (g/mol) __________________

Formula of hydrate ($M \cdot nH_2O$) ______ $M \cdot$ __ H_2O ______

The formula weight of the anhydrous salt of the unknown hydrates

Unknown #	Formula weight (g/mol)	Unknown #	Formula weight (g/mol)
1	120	13	120
2	32	14	213
3	64	15	120
4	71	16	284
5	96	17	192
6	96	18	64
7	142	19	120
8	128	20	71
9	120	21	120
10	96	22	32
11	160	23	64
12	96	24	71

Experiment 8   Heat of Fusion of Ice

Discussion

When a substance undergoes a change in its physical state, there must be an accompanying change in its energy. When a solid melts, for example, a definite amount of energy must be absorbed by the atoms, hold the particles together in the rigid positions of the solid. The quantity of energy required to overcome the attractive forces is dependent on the forces that exist between the particles.

The stronger the attractive forces in a substance, the larger the amount of energy that must be added in order to change the state of the substance. The purpose of this experiment is to measure the quantity of heat energy needed to overcome the crystal forces of attraction in ice. This quantity of heat energy is called the heat of fusion and is expressed as the amount of heat required to melt one gram of a solid at its melting point. A calorimeter will be used to determine this constant.

The principle of calorimetry is the Law of Conservation of Energy, which states that energy may be transferred from one object to another without loss or gain in the total energy involved. This means that in an isolated system (no energy is lost to, nor received from the surroundings) the loss in energy by one part of the system equals the gain in energy by the other part(s) of the system. The contents of a well–insulated calorimeter constitute such an isolated system; any energy lost by one part of the system is equal to the energy gained by the other part(s) of the system. One rarely attains the perfection of a situation in which no heat is lost. This is

especially true if liquids are involved, as they must be contained in and must be in intimate contact with a vessel that will absorb or release heat as the liquid changes temperature. The heat released or absorbed by the container is equal to the product of its heat capacity and the change in temperature. Thus the heat capacity of the container is the amount of heat that this particular container would release or absorb for each degree of change in temperature of its contents.

To work with heat transfers it is necessary to understand the meaning of the fundamental terms involved. These terms and their definitions are:

a. The specific heat of a substance is the ratio of the heat needed to raise the temperature of one gram of substance 1°C to that needed to raise the temperature of 1 gram of water 1°C.

b. A calorie is the quantity of heat needed to increase the temperature of 1 g of water by 1°C; 1 cal = 4.184 J.

c. The heat capacity of a body or an object is that quantity of heat needed to raise the temperature of the body or object 1°C.

d. Density is the mass per unit volume.

The Law of Conservation of Energy states that when ice is added to a sample of water, the heat lost by one part of the system is equal to the heat gained by another part of the system.

$$\text{Heat gained} = \text{Heat lost}$$

When this process occurs, the result will be water at a temperature intermediate between the temperature of the ice and the original water.

This means that heat is gained by the ice when it melts, and also by the water formed when the ice melts as it warms to the final temperature. Since the liquid and the calorimeter are assumed to be at the same temperature, this also means that the original water loses heat, and so does the calorimeter as they cool to the final temperature. When these ideas are incorporated into the equation, the following results:

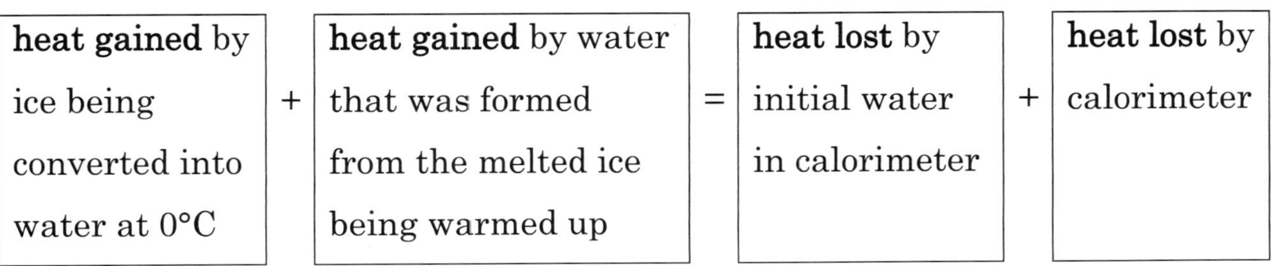

Thus, the amount of heat lost by the calorimeter must be determined. This heat will be equal to the product of the heat capacity of the calorimeter and the change in temperature. From this it can be seen that the heat capacity of the calorimeter must be determined before the measurement of the heat of fusion of ice.

Procedure

A. Determination of the heat capacity of the calorimeter

1. Place 100 g (weighed on a triple beam balance by difference) of water in the styrofoam calorimeter cup and record the temperature (t_1).

2. Place about 110 mL water in a beaker and heat to boiling. Assume the water temperature to be 100.0°C (t_1').

3. Wear a heat resistant glove and pour the boiling water to the water in the calorimeter. Immediately stir it with the thermometer and record the highest temperature reached (t_2). Measure the total mass of water

in the calorimeter. Subtract the original mass of water from the total mass of water to obtain the mass of boiling water added. Make two separate determinations of the heat capacity and average your results.

Calculations

a. Heat gained by cold water = $(m_{cw}) \times (c_w) \times (t_2 - t_1)$, where m_{cw} is the mass of cold water, c_w is the specific heat of water which is 4.184 J/(g·°C) (that is 1.00 cal/(g·°C), 1 cal = 4.184 J), t_2 is the final equilibrium temperature (°C), t_1 is the initial temperature of the cold water (°C);

b. Heat gained by calorimeter = $(HC_{cal}) \times (t_2 - t_1)$, where HC_{cal} is the heat capacity of the calorimeter;

c. Heat lost by hot water = $(m_{hw}) \times (c_w) \times (t_2 - t_1')$, where m_{hw} is the mass of hot water, t_1' is the initial temperature of the hot water (°C).

The Law of Conservation of energy says, heat gained = heat lost, therefore

$$(m_{cw}) \times (c_w) \times (t_2 - t_1) + (HC_{cal}) \times (t_2 - t_1) = -(m_{hw}) \times (c_w) \times (t_2 - t_1')$$

Solving for the heat capacity of the calorimeter gives HC_{cal}.

B. Determination of the heat of fusion of ice

1. Place 200 g of water in the calorimeter and record the temperature of the water (t_1).

2. Select several pieces of ice that weigh collectively between 15 and 25 grams. It is not necessary to weigh the ice very accurately at this time. Blot each lump to remove as much liquid water as possible and add the blotted lumps to the water in the calorimeter. Stir the ice/water mixture

until the ice has melted and record the lowest temperature obtained (t_2). Assume the original temperature of the ice to be 0.0°C (t_1').

3. Measure the total mass of water in the calorimeter as done previously in Part A. The difference in the total mass of water and the initial mass is the mass of water from the melted ice.

4. Make two separate determinations of the heat of fusion of ice and average your results. Use the previously determined heat capacity of your calorimeter in this calculation.

Calculations

a. Heat gained by melting ice = (mass of ice)×(Heat of fusion of ice) = $(m_{ice})×(\Delta H_{fusion})$;

b. Heat gained by water from the melted ice = (mass of ice)×(specific heat of water)×(Δt) = $(m_{ice})×(c_w)×(t_2 - t_1')$, where t_2 is the final equilibrium temperature (°C), and t_1' is the initial temperature of this water (that is 0°C from ice);

c. Heat lost by initial water = (mass of initial water)×(specific heat of water)×(Δt) = $(m_{water})×(c_w)×(t_2 - t_1)$, where t_2 is the final equilibrium temperature (°C), and t_1 is the initial temperature of this water (°C);

d. Heat lost by calorimeter = (heat capacity of calorimeter)×(Δt) = $(HC_{cal})×(t_2 - t_1)$.

From the Law of Conservation of Energy, heat gained = heat loss, hence

$$(m_{ice})×(\Delta H_{fusion}) + (m_{ice})×(c_w)×(t_2 - t_1')$$
$$= - [(m_{water})×(c_w)×(t_2 - t_1) + (HC_{cal})×(t_2 - t_1)]$$

Solving for the heat of fusion of ice, ΔH_{fusion}.

Lab report

In addition to the required contents, the following information should also be included in your lab report:

A. Determination of heat capacity of calorimeter

	Trial 1	Trial 2
Initial mass of cold water in calorimeter (g)	______	______
Temperature of the initial cold water (°C)	______	______
Total mass of water after mixing (g)	______	______
Mass of hot water after mixing (g)	______	______
Final temperature of the mixture (°C)	______	______
Heat capacity of calorimeter (J/°C)	______	______
Average heat capacity of calorimeter (J/°C)	______	

B. Determination of heat of fusion of ice

	Trial 1	Trial 2
Initial mass of water in calorimeter (g)	______	______
Initial temperature of water (°C)	______	______
Total mass of water after addition of ice (g)	______	______
Mass of ice (g)	______	______
Final temperature of the mixture (°C)	______	______
Heat of fusion of ice (J/g)	______	______
Average heat of fusion of ice (J/g)	______	

Experiment 9   Heat of Reaction

Discussion

Every chemical change is accompanied by a change in energy, usually in the form of heat. The quantity of energy released or absorbed during a chemical reaction is referred to as the heat of reaction. Chemical reactions that occur with absorption of heat are termed endothermic whereas those reactions which release heat are exothermic. Reactions can be conveniently studied in the laboratory under conditions of constant pressure. Under these conditions, the heat change in the system is referred to as the change in enthalpy, and is given the symbol ΔH. When heat is evolved in an exothermic reaction, ΔH is negative. When heat is absorbed in an endothermic reaction, ΔH is positive. Thus, for the reaction

$$2\ H_2 + O_2 = 2\ H_2O, \Delta H = -285.77\ kJ$$

The source of the negative sign for ΔH in exothermic reactions results from the change in enthalpy being defined as the difference between the enthalpy of the products and the enthalpy of the reactants

$$\Delta H = (\text{heat content of products}) - (\text{heat content of reactants})$$

This equation can be rewritten as:

$$\Delta H = \Sigma H(\text{products}) - \Sigma H(\text{reactants})$$

This experiment will investigate the heat of reaction of two types of reactions: (1) dissolution of a solid in water; (2) acid–base neutralization. The quantity of heat transferred will not be directly measured in the experiments. Instead temperature changes will be observed as indicators of the heat effects, and then the heat change will be calculated.

Dissolution of a solid

When a salt dissolves there are two major competing heat effects. The energy required to break the ionic forces of the solid is called the lattice energy. As the free ions become hydrated, the liberated energy is known as the hydration energy. If the lattice energy is greater than the hydration energy the reaction will be endothermic. If the hydration energy is greater than the lattice energy the reaction will be exothermic. In this experiment, we will not be concerned with measuring the lattice or hydration energies. Instead, we will deal with the overall heat changes when two ionic solids are dissolved in water.

Neutralization reactions

When hydrochloric acid is added to a sodium hydroxide solution, heat is evolved. Since the reaction of an acid with a base is called neutralization, the heat is called the heat of neutralization. The equation is:

$$HCl + NaOH = NaCl + H_2O + heat$$

Water is used as the solvent for the preparation of NaOH and HCl solutions, and therefore the acid, base, and salt are assumed to exist as ions in solution.

$$H^+ + Cl^- + Na^+ + OH^- = Na^+ + Cl^- + H_2O + heat$$

Sodium and chloride ions appear on both sides of the equation, so they can be subtracted out to give what is called a net ionic equation.

$$H^+ + OH^- = H_2O + heat$$

According to the net ionic equation, a neutralization reaction between a strong acid and a strong base is the reaction of hydrogen ions (H^+) and the hydroxide ions (OH^-) to form water. Since this reaction is common to all strong acids and bases, and since a definite amount of heat is liberated for each mole of reacting hydrogen ion and hydroxide ion, it is possible to use different strong acids with a fixed amount of base and obtain the same quantity of heat for each neutralization reaction. This, however, is beyond the scope of this experiment. The heat that is evolved in the neutralization reaction causes a rise in the temperature of the solution after the acid and base are mixed. Measurement of the temperature change, using measured volumes of known concentrations of NaOH and HCl, allows the heat of reaction to be calculated.

Procedure

A. Dissolution of ammonium nitrate (NH_4NO_3)

Obtain a styrofoam cup and rinse it thoroughly with water. Measure 100 mL of water in a graduated cylinder and pour it into the cup. Weigh approximately 5 ~ 7 grams of ammonium nitrate crystals and record the actual weight (to 0.01 g). Measure the water temperature every 30 seconds

for 2 minutes. Pour the NH_4NO_3 into the water and thoroughly mix to dissolve the solid. Continue stirring and record the temperature of the mixture at 30 second intervals for a period of 4 minutes after mixing. The temperature change is the lowest temperature recorded after mixing minus the initial temperature. Making a graph of temperature versus time will help you see the temperature change.

B. Dissolution of magnesium sulfate ($MgSO_4$)

Repeat the procedure for A again using 100 mL of water but this time use 4 ~ 6 grams of anhydrous magnesium sulfate. Be sure to record temperature measurements at 30 second intervals for 2 minutes before mixing and 4 minutes after mixing. This time the temperature change is the highest temperature after mixing minus the initial temperature. Make a graph of temperature versus time and plot your data.

C. Acid–base neutralization

Measure 50 mL of 2.0 M HCl solution with a graduated cylinder and transfer it to the empty calorimeter cup. Take 50 mL of 2.0 M NaOH solution and, making sure that it is at the same temperature as the HCl solution, pour it rapidly into the HCl solution with rapid stirring using the thermometer. Record the temperature of the solutions before mixing and the highest temperature after mixing. The amount of heat liberated can be found from the temperature change using the total volume of the solution after mixing and the specific heat of water. Express the heat of reaction in terms of kilojoules per mole of water formed. Repeat this experiment once.

Calculations

A. Dissolution of ammonium nitrate.

This reaction is endothermic, so the temperature will drop after mixing, and heat is absorbed by the solution.

Heat lost = Heat of reaction (heat of dissolution)

Heat lost = heat lost by solution + heat lost by calorimeter

Heat lost = $(m_w) \times (c_w) \times (\Delta t) + (HC_{cal}) \times (\Delta t)$, where $\Delta t = t_2 - t_1$, t_2 is the final temperature and t_1 is the initial temperature of water; the specific heat of water (c_w) has a value of 4.184 J/(g·°C), and the heat capacity of the calorimeter (HC_{cal}) is 58.6 J/°C.

Convert the mass of ammonium nitrate that was used in the experiment to moles, and then divide the heat loss value obtained from this equation by the number of moles to obtain the heat of reaction per mole of ammonium nitrate.

B. Dissolution of magnesium sulfate

This reaction is exothermic, so the temperature will rise after mixing, and heat is liberated by the reaction.

Heat of reaction = Heat gained

Heat gained = heat gained by solution + heat gained by calorimeter

Heat gained = $(m_w) \times (c_w) \times (\Delta t) + (HC_{cal}) \times (\Delta t)$, where $\Delta t = t_2 - t_1$, t_2 is the final temperature and t_1 is the initial temperature of water. Once again, the specific heat of water is 4.184 J/(g·°C), and the heat capacity of the calorimeter is 58.6 J/°C.

Convert the mass of magnesium sulfate dissolved in the experiment into moles of magnesium sulfate, and divide the value obtained from this equation by this number of moles in order to obtain the heat of reaction per mole of magnesium sulfate.

C. Heat of neutralization

Heat gained = Heat of neutralization

Heat gained = heat gained by solution + heat gained by calorimeter

Heat gained = $(m_w) \times (c_w) \times (\Delta t) + (HC_{cal}) \times (\Delta t)$, where $\Delta t = t_2 - t_1$, t_2 is the final temperature and t_1 is the initial temperature of the solution. Again, the specific heat of water is 4.184 J/(g·°C), and the heat capacity of the calorimeter is 58.6 J/°C.

Calculate the number of moles of water formed in the reaction, and divide the heat gained by this number to calculate the heat of reaction per mole of water formed.

Lab report

In addition to the required contents, the following information should also be included in your lab report:

A. Mass of ammonium nitrate (g)   ____________

    Moles of ammonium nitrate (mol)   ____________

    Volume of water (mL)   ____________

    Mass of water (g)   ____________

    Initial temperature (°C)   ____________

    Lowest temperature after mixing (°C)   ____________

Amount of heat absorbed (J)   __________

Heat per mole of ammonium nitrate (kJ/mol)   __________

B. Mass of magnesium sulfate (g)   __________

Moles of magnesium sulfate (mol)   __________

Volume of water (mL)   __________

Mass of water (g)   __________

Initial temperature (°C)   __________

Highest temperature after mixing (°C)   __________

Amount of heat liberated (J)   __________

Heat per mole magnesium sulfate (kJ/mol)   __________

	Trial 1	Trial 2
C. Volume of 2.0 M HCl (mL)	__________	__________
Volume of 2.0 M NaOH (mL)	__________	__________
Total volume (mL)	__________	__________
Moles of water formed (mol)	__________	__________
Initial temperature (°C)	__________	__________
Highest temperature (°C)	__________	__________
Heat liberated (J)	__________	__________
Heat per mole of water formed (kJ/mol)	__________	__________

Average heat per mole of water formed (kJ/mol)   __________

Discussion

Equal volumes of gases at the same temperature and pressure contain the same number of molecules. Thirty two grams of oxygen (O_2) has been experimentally shown to occupy 22.4 liters at STP (Standard Temperature and Pressure that are 0°C and 760 mmHg). Therefore, 1 mol of any gas occupies 22.4 liters at STP; this volume is therefore called gas molar volume at STP. In performing this experiment it is not practical to control the temperature and pressure at STP. Collecting a fraction of the molar volume at room temperature and atmospheric pressure and subsequent conversion to STP is sufficient.

In this experiment the oxygen is prepared from potassium chlorate that decomposes readily when heated, giving potassium chloride and oxygen. A small quantity of Fe_2O_3 serves as a catalyst. The loss in mass of the test tube containing the $KClO_3$ represents the mass of O_2 prepared.

In this experiment, Dalton's Law will be used to find the pressure of the oxygen gas generated. By the application of Boyle's and Charles' laws, the volume of oxygen gas generated will be converted to the corresponding volume at standard temperature and pressure. The underlying basis for the experiment is Avogadro's Law. The measurements in this experiment enable us to perform the following calculations, among others:

a.   The percentage by weight of oxygen in potassium chlorate;

b.   The volume of 1 mole of any gas at STP (the molar volume of gas);

c.   The density of oxygen gas at STP.

Procedure

1. Weigh an 8" test tube using an analytical balance to the nearest 0.1 mg. Add 0.7 ~ 0.8 g (using a top–loading balance) of $KClO_3$ to the test tube and weigh again to the nearest 0.1 mg. Add ~ 0.1 g Fe_2O_3 to the test tube and weigh again to the nearest 0.1 mg. Shake the test tube slightly to make the two solid powders well mixed.

2. Assemble the experimental apparatus as shown in Figure 15. Make tube A in the bottle close to the bottom. Fill water to the neck of the bottle and make the large rubber stopper a tight seal on the bottle.

3. Remove the small rubber stopper from the test tube, attach a rubber bulb on tube B to blow water through the tube A until the tube A is full of water to the nozzle. While water is still flowing through tube A tighten the pinch clamp and stop blowing. No more water should siphon over. Note: do not blow too much water to the beaker; the remained water level in the bottle should be close to the bottle shoulder. Otherwise refill the bottle and repeat this step.

4. Make a tight seal with the small rubber stopper as shown in Figure 15.

5. Open the pinch clamp, and allow water in the beaker to flow into the beaker as needed. If the system is air tight, water will cease to flow almost immediately.

6. Keep nozzle of the tube A (in the beaker) below the surface of the water and the clamp open, raise the beaker until the upper level of the water in the beaker is even with the upper level of the water in the bottle. This is to make the inner pressure the same as the room pressure. While the two levels are the same, close the pinch clamp. Discard the water in the beaker and weigh the dry beaker (to 0.1 g), then place it back to the position. Open the clamp; a small amount of water will

siphon out, which should be retained. Have your instructor approve the apparatus at this point if needed.

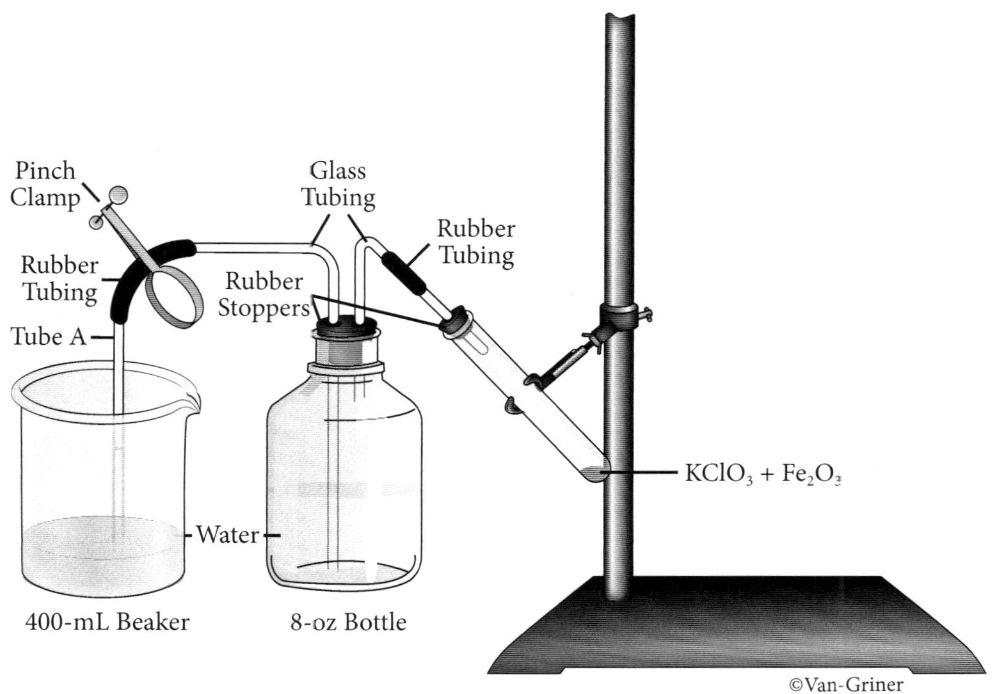

Figure 15. The experiment setup.

7. With the pinch clamp open, heat the test tube **gently** so as not to decompose the $KClO_3$ so rapidly that excessive pressure is built up in the system. This can be achieved by slowly moving the flame back and forth underneath the test tube contents. Continue the liberation of O_2 until no more oxygen is generated (i.e., no more water is driven over to the beaker). Keep nozzle of the tube A (in the beaker) below the level of the water and allow the test tube to cool. Some water will be drawn back into the bottle.

8. When the test tube have returned to room temperature, raise the beaker (or raise the bottle) until the two water upper levels are once again the same and close the clamp. Weigh the water in the beaker with

68

a triple beam balance to the nearest 0.1 g. Weigh the test tube to the nearest 0.1 mg and record the temperature and barometric pressure of the room (from the instructor).

Calculations

1. The mass of O_2 generated. When potassium chlorate is heated, it decomposes, forming solid potassium chloride and oxygen gas. Potassium chlorate decomposes slowly at 400°C, but with the presence of a catalyst, such as Fe_2O_3, it readily decomposes at about 270°C.

$$2 \, KClO_3 \, (s) = 2 \, KCl \, (s) + 3 \, O_2 \, (g)$$

Because gases are difficult to weigh directly, the weight of oxygen evolved is obtained indirectly by the difference of the test tube and contents before and after reaction:

$$m_{O2} = (m_{KClO3} + m_{Fe2O3}) - (m_{KCl} + m_{Fe2O3})$$
$$= (m_{test\,tube} + m_{KClO3} + m_{Fe2O3}) - (m_{test\,tube} + m_{KCl} + m_{Fe2O3})$$
$$= \text{total mass of the test tube and contents before heating} - \text{total}$$
$$\text{mass of the test tube and contents after heating}$$

These data enable one to calculate the percentage of oxygen in $KClO_3$:

$$\% \text{ of } O = (m_{O2} \, / \, m_{KClO3}) \times 100$$

2. The volume of O_2 generated. The oxygen was collected by displacing an equal volume of water from the bottle. From the mass of the displaced

water in the beaker, you can use the water's density to calculate the volume of the water that is equal to the volume of the generated oxygen gas.

3. The thus obtained volume of oxygen gas is the volume at the temperature and pressure in the laboratory, which has to be converted to the volume at STP. In making the correction to standard conditions, you must consider that the gas collected is saturated with water vapor. The total pressure of the gas is the sum of the partial pressures of oxygen gas and water vapor. Subtract the vapor pressure of water from the barometric pressure to get the actual partial pressure of the prepared oxygen under the experimental conditions: $P_{O2} = P_{atmosphere} - P_{water\ vapor}$.

It is now possible to calculate the volume the oxygen would occupy under conditions of STP (i.e., 760 mmHg pressure and 0°C) by use of the following:

$$\frac{P_1 V_1}{T_1} = \frac{P_2 V_2}{T_2}$$

where subscript 1 stands for the values obtained at experimental conditions (room temperature and atmospheric pressure), subscript 2 is for the values at STP. In other words, $P_1 = P_{O2}$ (mmHg), $V_1 = O_2$ volume obtained in the experiment (L), T_1 = room temperature (K), P_2 = 760 mmHg pressure, and $T_2 = 273.15$ K (0°C), V_2 is the volume occupied under STP conditions for the generated oxygen (L). This V_2 is the STP volume of oxygen you obtained in the experiment. From this volume and the mass of oxygen, you can calculate the volume occupied by the oxygen of 32.00 grams (1 mol).

4. The density of oxygen is calculated by division of the mass of oxygen produced by the volume that the oxygen occupies at STP.

Lab report

In addition to the required contents, the following information should also be included in your lab report:

Mass of oxygen produced (g) 26.25g

Mass of $KClO_3$ (g) 90.50g

% O in $KClO_3$ (%) 29%.

Mass of water driven over (g) 217.6g

Temperature of water (°C) 20.0°C

Density of water (see page 41) (g/mL) 0.998203g/mL

Volume of water driven over (L) 218L

Atmospheric pressure (mmHg) 769.50 mmHg

Vapor pressure of water

   (see the table next page) (mmHg) 17.552 mmHg

Partial pressure of O_2 (mmHg) 80mmHg

Volume of O_2, corrected to STP (L) 85L

Volume at STP occupied by 32 g of O_2 (L) 22.4 L

% error in molar volume of O_2 at STP ____

Density of O_2 at STP (g/L) 1.428g/L

Vapor Pressure of Water (mmHg) vs. Temperature (°C)

Temperature (°C)	Vapor Pressure (mmHg)	Temperature (°C)	Vapor Pressure (mmHg)	Temperature (°C)	Vapor Pressure (mmHg)
15.0	12.833	20.0	17.552	25.0	23.756
15.1	12.915	20.1	17.660	25.1	23.898
15.2	12.997	20.2	17.769	25.2	24.040
15.3	13.080	20.3	17.879	25.3	24.184
15.4	13.164	20.4	17.989	25.4	24.328
15.5	13.247	20.5	18.100	25.5	24.472
15.6	13.332	20.6	18.211	25.6	24.618
15.7	13.417	20.7	18.323	25.7	24.764
15.8	13.502	20.8	18.436	25.8	24.911
15.9	13.588	20.9	18.549	25.9	25.059
16.0	13.674	21.0	18.663	26.0	25.208
16.1	13.761	21.1	18.777	26.1	25.357
16.2	13.848	21.2	18.892	26.2	25.507
16.3	13.936	21.3	19.008	26.3	25.658
16.4	14.024	21.4	19.124	26.4	25.810
16.5	14.113	21.5	19.241	26.5	25.963
16.6	14.202	21.6	19.359	26.6	26.116
16.7	14.292	21.7	19.477	26.7	26.270
16.8	14.382	21.8	19.596	26.8	26.425
16.9	14.473	21.9	19.715	26.9	26.581
17.0	14.564	22.0	19.835	27.0	26.738
17.1	14.656	22.1	19.956	27.1	26.895
17.2	14.748	22.2	20.078	27.2	27.053
17.3	14.841	22.3	20.200	27.3	27.213
17.4	14.934	22.4	20.322	27.4	27.372
17.5	15.028	22.5	20.446	27.5	27.533
17.6	15.123	22.6	20.570	27.6	27.695
17.7	15.218	22.7	20.695	27.7	27.857
17.8	15.313	22.8	20.820	27.8	28.020
17.9	15.409	22.9	20.946	27.9	28.185
18.0	15.505	23.0	21.073	28.0	28.350
18.1	15.603	23.1	21.201	28.1	28.515
18.2	15.700	23.2	21.329	28.2	28.682
18.3	15.798	23.3	21.458	28.3	28.850
18.4	15.897	23.4	21.587	28.4	29.018
18.5	15.996	23.5	21.717	28.5	29.187
18.6	16.096	23.6	21.848	28.6	29.357
18.7	16.196	23.7	21.980	28.7	29.528
18.8	16.297	23.8	22.112	28.8	29.700
18.9	16.399	23.9	22.245	28.9	29.873
19.0	16.501	24.0	22.379	29.0	30.047
19.1	16.603	24.1	22.513	29.1	30.221
19.2	16.706	24.2	22.648	29.2	30.397
19.3	16.810	24.3	22.784	29.3	30.573
19.4	16.914	24.4	22.921	29.4	30.750
19.5	17.019	24.5	23.058	29.5	30.929
19.6	17.124	24.6	23.196	29.6	31.108
19.7	17.231	24.7	23.335	29.7	31.288
19.8	17.337	24.8	23.475	29.8	31.469
19.9	17.444	24.9	23.615	29.9	31.651

Experiment 11 Charles' Law

Discussion

Matter is able to expand when heated and contract when cooled. As you know, gases expand and contract to a much greater extent than do either solids or liquids. In this experiment you will investigate how the temperature of a gas influences its volume when the pressure remains unchanged. To do this, you will raise the temperature of a thin glass tube containing a trapped air sample, and then record volume changes as the air sample cools. Since the tubes have uniform diameters, the length of the tubes is proportional to the volume of the gas inside. It is not necessary, however, to determine the exact volume of the gas inside the tube, just its length. In this experiment you must assume that the air trapped in the tube and the thermometer are always at the same temperature.

Procedure

1. Fasten a glass capillary tube that is sealed at one end to the lower end of a thermometer using two bands of rubber tubing (Figure 16 top). The open end of the tube should be placed closest to the thermometer bulb, and ~ 5 mm above the end of the thermometer.

2. Immerse the tube and thermometer into a hot oil bath that has been prepared in the hood. Be sure that the entire length of the capillary tube is covered by the hot oil. Use caution since the oil is heated to about 120°C.

3. Wait for the tube and thermometer to reach the temperature of the oil bath (~ 120°C).

4. After the tube and thermometer have reached constant temperature, lift them so that about ¾ of the capillary tube is elevated out of the oil bath. Pause here for about 3 – 5 seconds to allow some of the oil to rise into the capillary tube.

5. **Quickly** carry the tube and thermometer (on a paper towel to avoid oil dripping) back to your desk. Lay the tube and thermometer on a paper towel on the desk. Make a reference line on the paper at the sealed end (top) of the capillary tube. Also mark the upper end of the oil plug (Figure 16 bottom). Alongside this mark write the temperature of the air column at this length.

6. As the temperature of the gas sample drops, make at least five marks representing the length of the air column trapped above the oil plug – measured from the end of the oil plug to the closed end of the tube. Allow enough time so that the temperature drops to 35 ~ 45°C. Since the tube has a uniform diameter, length serves as a relative measure of the gas volume.

7. Repeat steps 2 – 6 two additional times. You should use the same capillary tube setup for each of these two additional trials. Be sure to use different paper towels for your measurements of the additional trials.

8. Measure and record (in centimeters) the marked lengths of the gas sample during each trial. Remember that the length of the air column trapped in the tube is the distance from the top of the oil plug to the closed end of the tube.

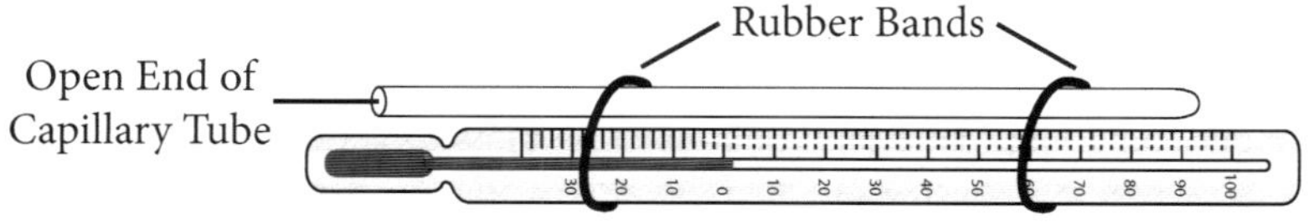

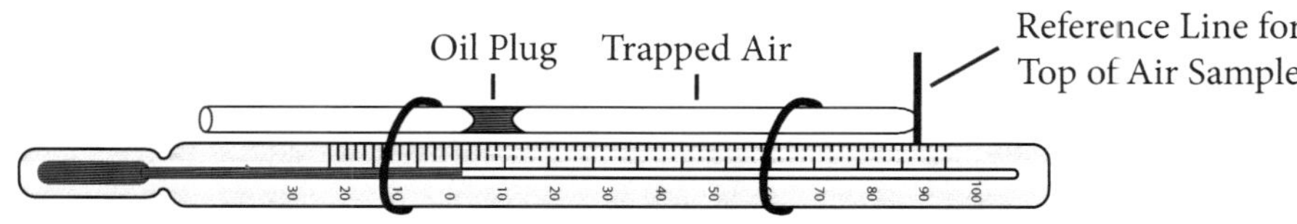

Figure 16. Charles' Law experiment setup.

Calculations

Prepare an Excel graph for each set of data (page 10, "Graphical Analysis by Excel Spreadsheets"). The length of the air column should be plotted on the y–axis, and the temperature of the air sample should be on the x–axis. Draw the best straight line through the data points, and find the temperature when length is 0. Or you can use the linear trendline function from Excel and calculate the temperature when length is 0 based on the linear regression line equation. Record this temperature.

Lab report

In addition to the required contents, the following information should also be included in your lab report:

Trial 1

Temperature (°C)							
Length (cm)							

Trial 2

Temperature (°C)							
Length (cm)							

Trial 3

Temperature (°C)							
Length (cm)							

For each trial, graph the length as the ordinate versus the temperature (you can use either °C or K) as the abscissa (refer to page 10 for "Graphical Analysis by Excel Spreadsheets"). Draw a straight line or use the linear regression to fit all the data points. Find or calculate the temperature when the length is 0 (zero).

	Trial 1	Trail 2	Trail 3
Temperature when length is 0:	__________	__________	__________

Average temperature when length is 0: __________.

Explain what this temperature means.

Experiment 12 Determination of Approximate Atomic Weight

Discussion

The purpose of this experiment is to determine the approximate atomic weight of several elements by means of calorimetry and the Dulong–Petit Law.

In 1819 P. Dulong and A. Petit discovered a relationship between the specific heat of certain metallic elements and their atomic weight:

$$\text{(specific heat)} \times \text{(atomic weight)} = 26.8$$

The specific heat of a substance is the ratio of the heat needed to raise the temperature of one gram of substance 1°C to that needed to raise the temperature of 1 gram of water 1°C. One calorie (4.184 J) is defined as the amount of heat necessary to raise the temperature of 1 gram of water 1°C.

It follows from the Dulong–Petit Law that atomic weight can be determined from specific heat measurements. In order to determine the specific heat of a substance it is necessary to measure the amount of heat required to change its temperature 1°C. This can be accomplished by measurement of the heat exchange from a hot metal in a calorimeter. The change in temperature of the calorimeter fluid allows determination of the specific heat of the metal and finally the atomic weight via the Dulong–Petit Law.

The Law of Conservation of Energy is utilized in this experiment, i.e., hot metal will impart heat energy to a calorimeter and its contents.

Procedure

The calorimeter used in this experiment is the same type as used in the "Heat of Fusion of Ice" experiment (page 53). The heat capacity that was determined for the calorimeter in that experiment will be suitable at this time. Add 50.0 g of water to the calorimeter and measure its temperature with a thermometer.

Weigh (using a top–loading balance) sufficient amount of a metallic element to fill a large test tube with 1 ~ 1½ inches of metal. Place the metal in the test tube. Fill a 400 mL beaker about four–fifths full with water and connect the test tube assembly as shown in the Figure 17.

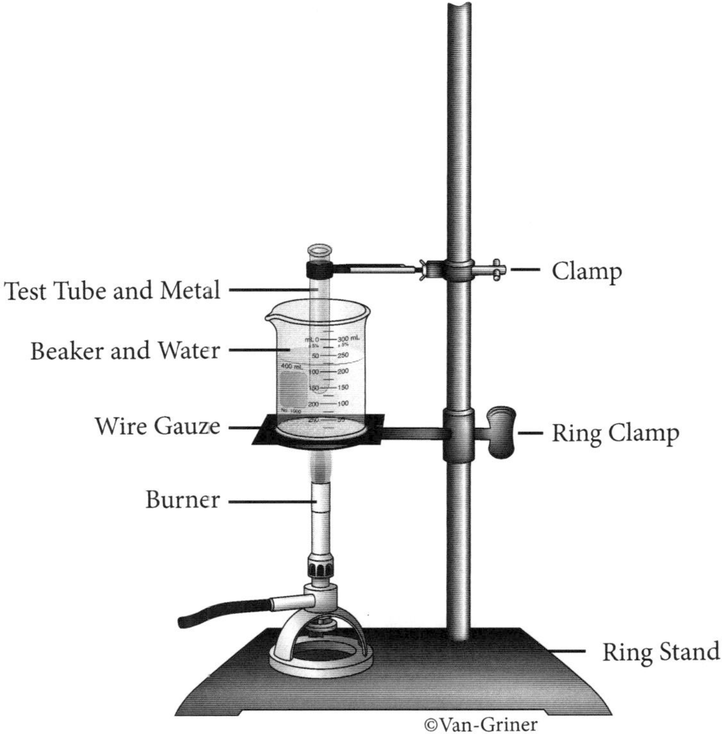

Figure 17. Atomic weight determination.

Heat the water in the beaker to boiling. Maintain the boiling water level several centimeters above the test tube contents for 10 ~ 15 minutes. Pour the metal rapidly into the water in the calorimeter and stir.

Record the maximum temperature of the mixture of metal and water. This temperature is used to calculate the heat loss of the metal as well as the heat gain of water. With the data calculate the approximate atomic weight of the element. Repeat the experiment to obtain reproducible values. Perform the experiment as often as time permits on the available metals. Compute the percent error based on exact weights and indicate results on your lab report.

Calculations

$$\text{Heat gain} = \text{Heat loss}$$

In this case the heat lost by the added metal will be gained by water and the calorimeter:

$$(m_{cw}) \times (c_w) \times (t_2 - t_1) + (HC_{cal}) \times (t_2 - t_1) = - (m_M) \times (c_M) \times (t_2 - t_1')$$

where m_{cw} is the mass of cold water, c_w is the specific heat of water which is 4.184 J/(g·°C), t_2 is the final equilibrium temperature (°C), t_1 is the initial temperature of the cold water (°C); HC_{cal} is the heat capacity of the calorimeter (58.6 J/°C); m_M is the mass of hot metal, c_M is the specific heat of the metal, t_1' is the initial temperature of the hot metal (assume to be 100.0°C).

The only unknown in the above expression is the specific heat of the metal (c_M). From this value and the equation of Dulong–Petit, the approximate atomic weight of the metal can be determined.

$$\text{(specific heat)} \times \text{(atomic weight)} = 26.8$$

$$\text{Atomic weight} = 26.8 \, / \, c_M$$

Lab report

In addition to the required contents, the following information should also be included in your lab report:

	Cu	Al	Zn
Mass of metal (g)	______	______	______
Initial temperature of water (°C)	______	______	______
Final temperature of water (°C)	______	______	______
Specific heat of metal (J/(g·°C))	______	______	______
Experiment–obtained atomic weight (g/mol)	______	______	______
True atomic weight (g/mol)	______	______	______
% error in atomic weight	______	______	______

Experiment 13   The Equivalent Weight of a Metal

Discussion

The purpose of this experiment is to determine the equivalent weight of a metal. Approximate atomic weight data from the previous experiment (Dulong–Petit) can be used in conjunction with the equivalent weight data from this experiment to determine an exact atomic weight of a metal.

The equivalent weight of an element is defined as that weight of an element that will react with or liberate 8.00 grams of oxygen or 1.00 gram of hydrogen. In this experiment aluminum metal will be allowed to react with HCl to produce hydrogen gas according to the reaction:

$$2\ Al\ (s) + 6\ HCl\ (aq) = 2\ AlCl_3 + 3\ H_2\ (g)$$

With the weight equivalence between aluminum and hydrogen gas, one can calculate the equivalent weight of aluminum.

Procedure

Assemble a molar volume apparatus as performed when oxygen was generated in the "Molar Volume of Gases" experiment (page 66). The only modification in this apparatus is a flexible rubber tube between the gas generation tube and the bottle with water. This allows regulation of the hydrogen gas production. Refer to Figure 15 on page 68.

Measure 10 mL of 6.0 M HCl solution and place it in the test tube. Weigh approximately 0.25 g (no more) of aluminum foil to ± 0.1 mg. Tie a piece of

thread around it, and attach it in the test tube shown in the figure. Be certain that the acid does not touch the metal until you are ready to begin the reaction. Adjust the apparatus for air leaks as before in the Molar Volume of Gases Experiment. Weigh a dry 600 ml beaker with a triple beam balance. When the reaction apparatus is ready, tilt the test tube to allow hydrochloric acid to make contact with the aluminum metal. Allow the reaction tube to cool and regulate the pressure inside the system to barometric pressure. Weigh the water displaced in the beaker. Measure the temperature of displaced water and the barometric pressure.

Calculations

From the volume of water displaced, barometric pressure, temperature of water, and vapor pressure of water at the measured temperature, calculate the volume of hydrogen gas under dry and STP conditions. Assume that one mole of hydrogen gas (2.016 g) occupies 22.4 liters at STP and calculate the grams of hydrogen gas produced. From the grams of aluminum reacted, calculate the equivalence between 1.008 grams of hydrogen and grams of aluminum. This will be the equivalent weight of aluminum.

From the experiment of the Determination of Approximate Atomic Weights (Dulong and Petit, Experiment 12, page 78), use the value of the atomic weight obtained for aluminum and the equivalent weight value obtained today and calculate the exact atomic weight of aluminum.

Lab report

In addition to the required contents, the following information should also be included in your lab report:

Weight of Al (g)   ______________________

Barometric pressure (mmHg)   ______________________

Room temperature (°C)   ______________________

Vapor pressure of water (mmHg)   ______________________

Volume of H_2 at experiment conditions (mL)   ______________________

Volume of H_2 at STP (mL)   ______________________

Weight of H_2 (g)   ______________________

Equivalent atomic weight of Al (g/mol)   ______________________

Approximate atomic weight of Al from Exp. 12 (g/mol)______________________

Exact atomic weight of Al (g/mol)   ______________________

Experiment 14

Determination of the Molecular Weight by Freezing Point Depression

Discussion

The vapor pressure of a liquid is a measure of the amount of that liquid that exists as a gas above the surface of the liquid. If a liquid has a high vapor pressure, then there is a large amount of vapor above the liquid. This will result in a liquid that will evaporate easily, and thus the liquid is termed volatile. Liquids with low vapor pressures have only a small amount of vapor above the surface of the liquid; they evaporate slowly, and are called nonvolatile.

The addition of a solute to a solvent will lower the vapor pressure of the solvent. This vapor pressure lowering of the solvent is related to observed colligative properties of solutions such as freezing point lowering, boiling point elevation and osmotic pressure.

This experiment is concerned with the freezing point lowering. A colligative property such as this is a property of the solute that depends only on the number of particles of solute in solution and not on their chemical characteristics. The more solute you add to a given mass of solvent, the more its freezing point will decrease. For example, one mole of sucrose dissolved in 1000 g of water will lower the freezing point of the water twice as much as one half of a mole of sucrose in 1000 g of water. In general, one mole of particles (6.022×10^{23}) of a solute will lower the freezing point of 1000 g of solvent the same amount. The numerical value of this lowering is a constant that is different for each solvent. Changing the number of

particles in a solution changes the concentration. A mathematical expression relating these quantities is given by:

$$\Delta T = K\,C$$

where ΔT = freezing point lowering of the solvent with a unit of °C;

    K = freezing point constant (temperature lowering of solvent with one mole of solute per 1000 g solvent) with an SI unit of kg·°C/mol;

    C = mass molar concentration of the solution with an SI unit of mol/kg.

The ΔT is the freezing temperature of the pure solvent minus the freezing temperature of the solvent with the solute added. The concentration of the solute in the solution depends on the number of moles of solute. The number of moles of solute is, in turn, calculated by dividing the mass of the solute by its molecular weight. As a result, the mass molar concentration of a solution can be related to the molecular weight of the solute.

$$C = \frac{\text{moles of solute}}{\text{mass of solvent}} \times 1000$$

$$= \frac{\text{mass of solute}}{\text{molecular weight of solute} \times \text{mass of solvent}} \times 1000$$

We use 1000 because the SI unit for mass is kg.

Now ΔT can be expressed as

$$\Delta T = K \times \frac{\text{grams of solute}}{\text{molecular weight} \times \text{grams of solvent}} \times 1000$$

This can be rearranged to give a direct calculation of the molecular weight of a nonvolatile solute:

$$\text{Molecular weight of solute} = K \times \frac{\text{grams of solute}}{\Delta T \times \text{grams of solvent}} \times 1000$$

Here is an example: 34.2 grams of sugar when dissolved in 1000 g of water lowered the freezing point of water from 0°C to −0.186°C. The freezing point constant for water is 1.86. The molecular weight of the sugar is:

$$\text{Molecular weight} = 1.86 \times \frac{(34.2 \text{ g sugar})}{[0 - (-0.186)] \times (1000 \text{ g water})} \times 1000 = 342 \text{ g/mol}$$

A discussion by the instructor of the experimental setup and nature of the cooling curve will be given. Special emphasis will be placed on **preventing supercooling**. Figures 18 and 19 illustrate the setup and cooling curve, respectively.

Procedure

1. A 600 ml beaker about two–thirds full of water will serve as the hot bath. Weigh 25.00 g of *p*–dichlorobenzene and transfer to a large test tube. Clamp the test tube in the water bath as shown in Figure 18 and begin heating. After most of the solid has melted, place the thermometer and the wire stirrer into the test tube. When all the

p–dichlorobenzene has melted, quickly remove the test tube from the bath, wipe the outside with a towel and place the test tube in a wide mouth bottle. While continuously stirring, record the temperature at one minute intervals. When the first crystals appear, the freezing point of *p*–dichlorobenzene has been reached and the temperature should level and remain constant until all the liquid has solidified. Construct a plot of temperature vs. time and determine the freezing point of *p*–dichlorobenzene.

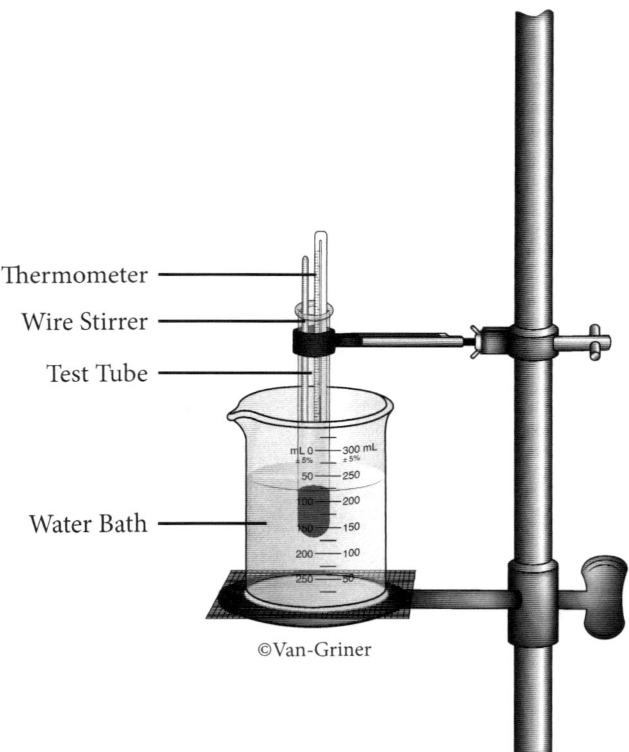

Figure 18. Experiment setup.

2. In a vial, obtain a sample of unknown solute. Weigh the vial with its contents. Empty about 1/3 of the unknown into the tube containing the solvent dichlorobenzene and then weigh again. This is solution 1. Set

the test tube in the hot bath. When all of the mixture has melted, stir thoroughly in order to ensure a homogeneous mixture. Once again transfer the test tube to a wide mouth bottle, and while continuously stirring record the temperature every minute.

3. Determine the freezing point of the solution. Empty the remaining contents of the vial into the test tube and weigh the vial. Repeat the above procedure to obtain the freezing point of the new mixture. This is solution 2.

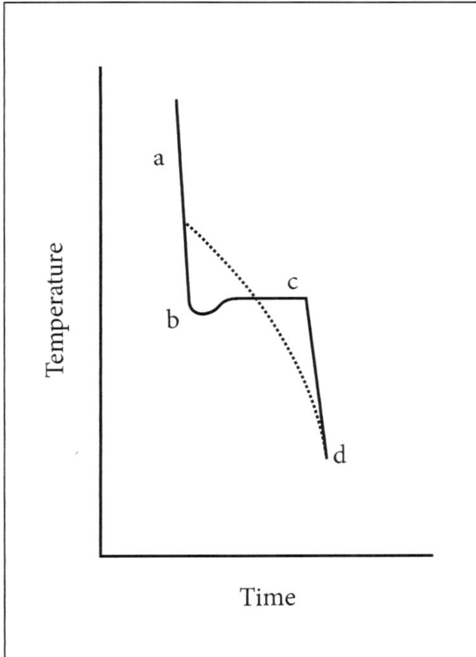

Figure 19. Cooling curve.

Solid line – slow cooling, good mixing; dotted line – if cooled too fast.

a – cooling of solution;

b – supercooling;

c – freezing point (liquid–solid);

d – cooling of solid.

Note: if solution is cooled too fast, the c portion of the curve may be missed.

Calculations

Subtract the freezing point of the two mixtures of different concentration from the freezing point of the pure dichlorobenzene. On a piece of graph

paper, plot these two ΔT values on the y–axis and the corresponding grams of solute in the mixture on the x–axis. Construct a straight line through these two points and the origin. Determine the molecular weight of the unknown by use of the second equation on page 86. The K for *p*–dichlorobenzene is 7.10 kg·°C/mol.

Lab report

In addition to the required contents, the following information should also be included in your lab report:

Freezing point of dichlorobenzene solvent (°C)   _______________

Freezing point of solution 1 (°C)   _______________

Freezing point of solution 2 (°C)   _______________

Weight of unknown solute in solution 1 (g)   _______________

Weight of unknown solute in solution 2 (g)   _______________

Weight of solvent dichlorobenzene (g)   _______________

Molecular weight of unknown in solution 1 (g/mol)   _______________

Molecular weight of unknown in solution 2 (g/mol)   _______________

Average molecular weight of unknown (g/mol)   _______________

Experiment 15   Preparation of Iron Sulfate

Discussion

Some compounds occur in nature as minerals or in common rocks, such as SiO_2 (sand), $NaCl$ (halite), and $CaCO_3$ (limestone). However, many other compounds are not found naturally and must be prepared from commonly occurring substances by means of chemical reactions. One method of synthesis allows a metal to react with an aqueous solution of a strong acid. For example,

$$Zn\ (s) + 2\ H_3O^+\ (aq) = Zn^{2+}\ (aq) + 2\ H_2O\ (l) + H_2\ (g)$$

Note that hydrogen gas is also generated. If the acid is hydrochloric, a solution of zinc chloride results:

$$Zn\ (s) + 2\ H_3O^+\ (aq) + 2\ Cl^-\ (aq) = Zn^{2+}\ (aq) + 2\ Cl^-\ (aq) + 2\ H_2O\ (l) + H_2\ (g)$$

Evaporation of the water would produce the solid salt, $ZnCl_2$. One or more molecules of water may remain with the salt as water of hydration (see Experiment 7). In this experiment, iron metal will react with sulfuric acid to make an iron sulfate.

Procedure

Weigh a 125 mL Erlenmeyer flask to the nearest 0.01 g using a top–loading balance. Add approximately 3.00 g of iron powder to the flask and reweigh. Label the flask with your initials and place it in the hood. Carefully add 25

mL of 3 M H_2SO_4. **DANGER!!** Sulfuric acid is a strong acid. **If spills occur, flood the area with cold water and notify the instructor.**

After the reaction flask is placed in the hood, prepare a water bath by placing 200 mL of tap water in a 400 mL beaker containing 4 or 5 boiling stones. Heat the water to boiling using a Bunsen burner. When the water begins to boil, remove the burner and extinguish the flame. Carefully place the hot water bath in the hood and insert your reaction flask into the hot water for a period of 20 ~ 30 minutes.

Immediately after removal from the water bath, gravity filters the contents of the flask to remove any unreacted iron powder. Collect the filtrate in a clean Erlenmeyer flask. Cool this flask and its contents in an ice bath for 15 minutes whereupon crystals should form. If crystallization does not result, ask the instructor to supply a seed crystal.

A suction filtration apparatus will be provided. Filter the crystals and wash the solid precipitate with 5 mL of ethyl alcohol. After filtration, spread the crystals out on a labeled filter paper that has been weighed to the nearest 0.01 g on the top−loading balance. Allow the crystals to dry in this manner.

When the crystals have dried, determine the weight of the product. Calculate the theoretical yield and percentage yield assuming that the product might be either of the following compounds: $Fe_2(SO_4)_3 \cdot 9H_2O$ or $FeSO_4 \cdot 7H_2O$.

Conduct Experiment 7 as found on page 49, Procedure B. Instead of an unknown, use about 2 g of the dry iron sulfate prepared above. Determine the formula of this hydrated iron sulfate according to the directions outlined, and use the report sheet provided on page 52 for calculation. Calculate two formulas for the iron sulfate hydrate prepared assuming that either $Fe_2(SO_4)_3 \cdot 9H_2O$ or $FeSO_4 \cdot 7H_2O$ might be present. Decide from the data which compound containing iron is most likely to be present.

Lab report

In addition to the required contents, the following information should also be included in your lab report:

Preparation of iron sulfate

Mass of iron + weighing paper (g)   ______________________

Mass of weighing paper (g)   ______________________

Mass of iron sulfate (g)   ______________________

Theoretical yield calculations

Iron (III) sulfate nonahydrate, $Fe_2(SO_4)_3 \cdot 9H_2O$, F.W. = 561.91 g/mol,

anhydrous F.W. = 399.91 g/mol   ______________________

Iron (II) sulfate heptahydrate, $FeSO_4 \cdot 7H_2O$, F.W. = 277.92 g/mol,

anhydrous F.W. = 151.92 g/mol   ______________________

Percent actual yield calculations:

Iron (III) sulfate nonahydrate, $Fe_2(SO_4)_3 \cdot 9H_2O$ ______________________

Iron (II) sulfate heptahydrate, $FeSO_4 \cdot 7H_2O$ ______________________